Bibliografische Information Der Deutschen Bibliothek

Die Deutsche Bibliothek verzeichnet diese Publikation in der Deutschen Nationalbibliografie; detaillierte bibliografische Daten sind im Internet über http://dnb.ddb.de abrufbar.

ISBN 3-8325-1203-9

Logos Verlag Berlin
Comeniushof, Gubener Str. 47,
10243 Berlin
Tel.: +49 030 42 85 10 90
Fax: +49 030 42 85 10 92
INTERNET: http://www.logos-verlag.de

Food quality and food choice in freshwater gastropods:

Field and laboratory investigations on a key component of littoral food webs

Patrick Fink

Logos Verlag Berlin 2006

Titelillustration: E. Fink 2005

Dissertation der Universität Konstanz

Tag der mündlichen Prüfung: 21. Oktober 2005

Referent: Prof. Dr. K.-O. Rothhaupt

Referent: PD Dr. E. von Elert

To Margot Siegert

(7.9.1926 – 3.5.2005)

Chapter 1

General Introduction

Herbivorous gastropods as important trophic links in littoral ecosystems

Analogous to cladoceran zooplankton in planktonic systems, gastropod grazers are a key group of herbivores at the plant-herbivore interface in many marine and freshwater benthic systems (Steinman 1996, Anderson and Underwood 1997). In Lake Constance, but also in many other lakes and rivers in the temperate zones of Europe, *Bithynia tentaculata* and *Radix ovata* are among the dominant gastropod species (Brendelberger 1997a, Baumgärtner 2004).

B. tentaculata is a prosobranch species and is able to switch between two feeding modes. With its gills, it can filter-feed on plankton, but it can also ingest periphyton which is scraped off the surface of hard substrates with the radula (Brendelberger and Jürgens 1993). In the feeding mode of radular scraping, *B. tentaculata* is able to feed selectively on high quality food (Brendelberger 1995b). This combination of high selectivity and the ability to switch feeding modes probably makes it a highly competitive species (Brendelberger 1997a).

R. ovata belongs to the pulmonate snails, is a secondary aquatic species and has some relevance in public health issues. It is one of the most important intermediate hosts for the trematode *Trichobilharzia* ssp., the parasitic agent of the so-called „swimmers itch". The members of the genus *Radix* are primarily herbivorous periphyton grazers (Calow 1970, Lodge 1986). Adult *R. ovata* reach shell lengths of up to 25 mm, which results in a rather high individual grazing impact. The combination of high abundance (Baumgärtner 2004) and high individual grazing impact makes it to a keystone link in the littoral food web, making primary production available to invertebrate (crayfish) and vertebrate (fish, waterfowl) predators (Turner et al. 2000). Thus, it can be assumed that constraints of the biomass accrual of gastropod grazers can also affect other trophic levels and thereby have high impact on the structure of freshwater littoral food webs (Liess and Hillebrand 2004).

Food quality for aquatic herbivores

The plant-herbivore interface constitutes the crucial step for the transfer of photosynthetically fixed solar energy to higher trophic levels in food webs. To understand how this step is regulated, it is essential to determine the factors that control the efficiency of this transfer of energy, which is known to be highly variable. Often, this variability can be attributed to the primary producers' nutritional quality for herbivore consumers. This nutritional quality can be measured as the resulting biomass accrual or reproductive output per unit of consumed resource (e.g. carbon). However, constraints of food quality and food quantity are often difficult to separate (Sterner 1997, Sterner and Schulz 1998).

Especially in freshwater plankton, multiple factors determining food quality of algae and cyanobacteria for herbivores have been examined (Sterner and Schulz 1998). One of these factors is the morphology of the algal or cyanobacterial cells. For example, large cyanobacterial filaments interfere with the food-gathering process of filter-feeding cladocera (DeMott et al. 2001), and green algae can resist digestion in the gut of the herbivores due to thick cell walls (Van Donk and Hessen 1993) or mucilaginous sheaths (Porter 1975). However, not only the morphology is important, but also the compounds within the cells have to be considered in the context of food quality. Various taxa of eukaryotic algae and especially cyanobacteria produce toxins that greatly reduce their nutritional value for herbivores and may even kill the latter (Lampert 1981). Furthermore, the lack or low availability of essential compounds can constrain the quality of ingested food particles. Such essential compounds could be mineral nutrients (e.g., nitrogen (N) and phosphorous (P), Sterner and Elser 2002) as non-substitutable resources, as well as essential organic compounds, such as amino acids or particular lipid compounds which the herbivore cannot synthesize in sufficient amounts by itself (e.g., Raubenheimer 1992, Von Elert 2002, Von Elert et al. 2003).

The role of N and P availability for the growth of homeostatic organisms

Among the non-substitutable elemental nutrients, especially nitrogen (N) and phosphorous (P) have gained ecologists' attention, because their limited and varying availability is known to lead to highly variable C:N:P ratios in primary producers (Elser et al. 2000b). In contrast to the high variation of their algal diet, herbivorous zooplankton maintain homeostatic, i.e. relatively constant body C:N:P ratios (Hessen

1990, Hessen and Lyche 1991). There is accumulating evidence from a broad range of ecosystems (ranging from terrestrial to marine and freshwater systems) that high C:N and C:P ratios in primary producers may result in insufficient availability of N and P for homeostatic herbivores and thus lead to limitation of herbivore growth (Urabe and Watanabe 1992, Sterner 1993, Elser et al. 2000b).

However, the studies in fresh waters focussed almost exclusively on interactions between planktonic organisms. Despite the high productivity of littoral algal communities (Pinckney and Zingmark 1993), studies on the nutrient status and elemental composition of periphyton are scarce (Kahlert 1998). This lack of data is even more pronounced for benthic herbivores which led Frost et al. (2002b) to the statement that „the first research priority" would be to assess the elemental composition of different benthic invertebrates .

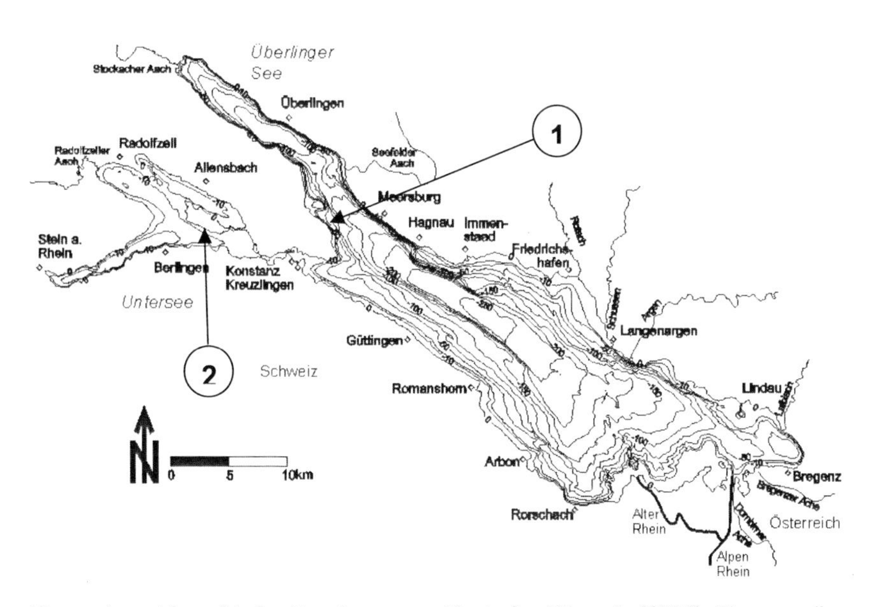

Figure 1: Map of Lake Constance, modified after Wessels (1998). The sampling sites of the field sampling programme are marked by the arrows; **1**) Litoralgarten; **2**) Reichenau.

I hypothesized that, similar to herbivorous zooplankton (Hessen and Lyche 1991), benthic herbivores in lake littorals are homeostatic in their body C:N:P ratios. In an initial step in order to estimate whether the variability of periphyton nutrient composition is higher than that of the herbivores, I investigated patterns of C:N:P ratios in the littoral of Lake Constance. **Chapter 2** describes the results of a field survey during the 2003 growth season. Two sites at Lake Constance (Fig. 1) were sampled monthly to assess temporal and spatial variation of the C:N:P ratios of both, the herbivorous macroinvertebrates and their periphyton resource. This is the first study ever that simultaneously investigated the C:N:P stoichiometry of consumer and resource at the littoral periphyton-herbivore interface.

However, neither high variability in periphyton C:N:P ratios, nor homeostasis of benthic consumers automatically implies an ecological relevance of these differences in variability. Thus, it remains to be tested if a „dietary mismatch“ has any consequences for the fitness of the herbivorous invertebrates. To date, only one study each in lentic (Frost and Elser 2002a) and lotic freshwater environments (Stelzer and Lamberti 2002) investigated the effects of low mineral nutrient availability on the growth of benthic invertebrates. Therefore, further investigations on the validity of the concept of 'Biological Stoichiometry' (sensu Elser et al. 2000a) for benthic food webs are necessary (Frost et al. 2002b).

In **chapter 3**, I investigated whether a limited availability of N and/or P has a negative impact on the growth of juvenile gastropods. Further, the interacting effects (Sterner and Schulz 1998) of food quantity (total C available) and quality (differences in C:N and C:P ratio) were tested. A growth experiment with juvenile *R. ovata* was performed in the laboratory by crossing different stoichiometries of the algal food with high and low food quantities,. Not only growth, but also food consumption and faeces of the snails were analyzed in order to account for possibly increased nutrient assimilation and for compensatory feeding as mechanisms to cope with low availability of mineral nutrients in the food source. Field data of *R. ovata* C:N:P ratios were collected for comparison with the laboratory results.

Biochemical resources - polyunsaturated fatty acids

Recently, it was discovered, that despite sufficient availability of elemental nutrients such as N and P, growth of organisms can be limited by the availability of biochemical resources, such as lipids (e.g., Wacker and von Elert 2001, Von Elert et al. 2003). Especially the absence of long-chain polyunsaturated fatty acids (PUFAs) has been proposed to be a potential food-quality constraint in zooplankton (Müller-Navarra 1995, Wacker and von Elert 2001). Especially PUFAs of the n-3 class cannot be synthesized by most invertebrates (Stanley-Samuelson et al. 1988). Therefore, these organisms have to acquire the necessary n-3 PUFAs with their diet. Among these n-3 PUFAs, α-linolenic acid (C18:3 n-3, Fig. 2A) and eicosapentaenoic acid (C20:5 n-3, Fig. 2B) seem to be of particular importance for the growth of zooplankton like representatives of the genus *Daphnia* (Müller-Navarra 1995, Wacker and von Elert 2001).

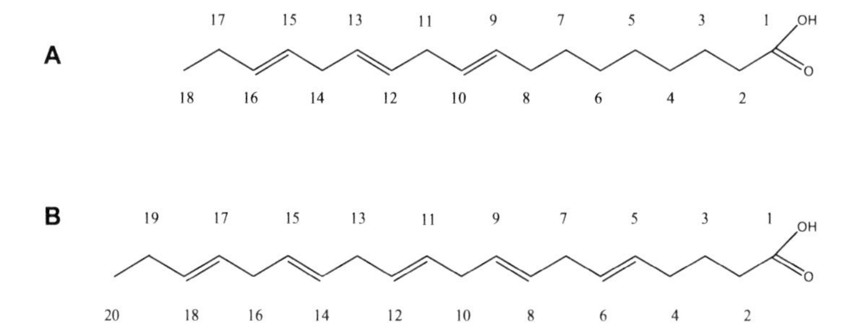

Figure 2: The nutritionally important n-3 PUFAs **A** α-linolenic acid C18:3 (9, 12, 15) and **B** eicosapentaenoic acid C20:5 (5, 8, 11, 14, 17)

However, studies regarding the role of PUFAs were so far almost exclusively performed for marine ecosystems (Pillsbury 1985, Delaunay et al. 1993) or for freshwater plankton (Müller-Navarra 1995, Wacker and von Elert 2001). Relatively little is known about the role of PUFAs in determining food quality effects in freshwater benthic organisms. Only recently, thorough investigations on molluscs have been performed. These have demonstrated a potentially limiting role of PUFAs in the determination of food quality for benthic herbivores, as the PUFA content of algae affected different life history stages of the bivalve *Dreissena polymorpha* (Wacker and von Elert 2002, Wacker and von Elert 2003, Wacker and von Elert 2004). Another group of molluscs, the gastropods, are important benthic herbivores in many freshwater littoral habitats (Feminella and Hawkins 1995), but have so far been neglected with respect to their susceptibility to limitation by a low availability of PUFAs.

In order to investigate the potential role of PUFAs in determining food quality of primary producers for freshwater gastropods, two laboratory growth experiments were conducted with the two (in terms of biomass) most abundant gastropod species from Lake Constance, the prosobranch *Bithynia tentaculata* (LINNAEUS) and the pulmonate Radix ovata (DRAPARNAUD). To disentangle effects of the food's PUFA composition and other determinants of food quality, the recently developed method of fatty acid supplementation (Von Elert 2002) was used to enrich food algae with single PUFAs. Thereby, algal suspensions were created that differed from the unsupplemented algae only in the content of a specific PUFA. These suspensions were subsequently used as food for juvenile snails in the growth experiments. In **chapter 4**, I depict a growth experiment with juvenile *B. tentaculata* in which the n-3 PUFAs α-linolenic acid and eicosapentaenoic acid were added as single compounds or in combination to a cyanobacterium and two species of freshwater planktonic algae. Juvenile growth rates, determined as increase in shell length as well as gain in soft body dry mass, were considered as indicators of food quality. In contrast to *B. tentaculata*, which can feed on both plankton and periphyton, the pulmonate gastropod *Radix ovata* is exclusively benthivorous. Thus, I performed a growth experiment with *R. ovata* fed on pure cultures of benthic algal strains with natural and experimentally modified PUFA patterns. This study, which is presented in **chapter 5**, focussed not only on the effects of PUFA availability towards the growth of the snails, but also on the impact of differing PUFA contents in the food organisms for the biochemical composition of the grazers themselves.

How to cope with variable qualities of the periphyton resource?

Periphyton is a highly variable food source. Not only biomass and taxonomic composition vary in both space and time (e.g., Rosemond et al. 2000, Kahlert et al. 2002), but also the algal stoichiometry is subject to constant variation (Frost et al. 2002a, Fink et al. submitted). This implies various potential constraints for the growth of benthic herbivores in lake littoral ecosystems. Hence, physiological or behavioural mechanisms that allow the herbivores to overcome (or at least reduce) growth limitations caused by low nutritional quality should be highly adaptive. Several such mechanisms have been suggested (Sterner and Hessen 1994).

For the herbivores, one way to cope with the low quality of a diet due to its low contents of essential compounds could be an enhanced consumption of food to gain access to sufficient amounts of the limiting nutrient; such active compensatory feeding has been demonstrated in a variety of organisms and ecosystems

(Raubenheimer 1992, Cruz-Rivera and Hay 2000), but so far it has not been explicitly demonstrated in the context of biological stoichiometry. Furthermore, organisms struggling with limited food quality could also increase the retention of the limiting dietary constituent during the digestive process. For example, animals limited by the availability of P could enhance their P retention in the gut and release excreta that are depleted in P relative to the food source. Excretion of high C:P faeces could also constitute a mechanism to get rid of excess carbon acquired during compensatory feeding. These two mechanisms were investigated in detail in an experiment with *R. ovata* and the results are presented in **chapter 3**.

As food availability for benthic herbivores is patchily distributed, food quality is probably patchy too. Thus, using information-transmitting chemical cues to detect patches of (high quality) food should be highly adaptive for these herbivores. Such foraging-kairomones (sensu Ruther et al. 2002) will be of particularly high value for organisms such as snails which have rather low abilities for visual orientation (Gal et al. 2004) and for which locomotion is associated with high energetic costs, primarily due to the production of pedal mucus (Denny 1980). It is well known that gastropods have strong chemotactic abilities which they use for various purposes such as trail following and mate choice (Croll 1983). Furthermore, food preferences have been extensively studied in various snail species from different habitats (e.g. Speiser and Rowell-Rahier 1991, Brendelberger 1995b, Wakefield and Murray 1998). However, it is unclear, in how far the mechanisms of food preferences and chemotaxis towards dissolved cues are connected. In particular, it is not known, if freshwater gastropods (like terrestrial snails, Croll and Chase 1980) are able to use foraging kairomones or if the mechanism of food-finding is rather determined by a 'random search' behaviour (Streit 1981).

Volatile organic compounds (VOCs) produced by algae and cyanobacteria are a frequent nuisance in water treatment, especially for drinking water. Musty and earthy smells, frequently attributed to cyanobacterial blooms cause severe economical costs as the removal of these odour compounds is difficult and expensive. On the other hand, the biological function of those odour compounds in aquatic environments is largely unknown. In terrestrial ecosystems, VOCs are known to be used as foraging kairomones by numerous species of insect herbivores in terrestrial ecosystems (Metcalf 1987). This led Watson (2003) to the hypothesis that similar to terrestrial ecosystems, VOCs could also function as infochemicals in aquatic habitats. And indeed, VOCs have been described to function as habitat-finding cues for both aquatic insects (Evans 1982) and nematodes (Höckelmann et al. 2004).

As outlined above, it should be highly adaptive for snails to be highly sensitive for chemical cues that allow to minimize the costs of food-finding by directed chemotaxis towards potential food sources. In **chapter 6**, I addressed the question whether gastropods are able to use algal volatiles as information transmitting cues indicating the presence and direction of algal food patches. The components of the bouquet of VOCs released upon cell damage from the benthic filamentous green alga *Ulothrix fimbriata* were determined by gas chromatography with mass spectrometry (GC/MS). An extract of *U. fimbriata* VOCs was tested for its attractivity to *R. ovata* in a specially designed new bioassay allowing the investigation of the behavioural response of aquatic gastropods to volatile infochemicals.

Chapter 2

Stoichiometric mismatch between littoral invertebrates and their periphyton food

Patrick Fink, Lars Peters, and Eric von Elert

Abstract

Ecological stoichiometry is considered a key concept in understanding constraints in energy transfer at the plant–herbivore interface. However, whether this concept is relevant for benthic freshwater ecosystems is not fully known. Therefore, a field survey was conducted in 2003 during the growing season in the littoral zone of Lake Constance, a large pre-alpine lake in central Europe. The aim was to assess temporal variation in the elemental stoichiometric composition in both herbivorous macroinvertebrates and their food resource, the periphyton in two different lakes. The periphyton showed large temporal and spatial variation in carbon, nitrogen, and phosphorus content, with particularly high molar C:P ratios of up to 1225:1. Periphyton C:P and C:N ratios were often high and constantly above the Redfield ratio that is considered optimal for autotrophic growth. In contrast to the pronounced fluctuations in the nutrient ratios of their food resource, the herbivorous macroinvertebrates showed only very little variation in their nutrient ratios, which indicated that they are homeostatic, i.e., physiologically restricted to a comparatively narrow range of C:P and C:N ratios. Distinct species-specific C:P and C:N ratios were found for different taxonomic groups of macroinvertebrates, which indicated different requirements of optimal dietary C:P and C:N ratios and which might influence the ability of the taxa to compete for limiting elemental nutrients. Considering the temporally very high C:P and C:N ratios of the periphytic resource and the very low ratios of the consumer body tissue, this stoichiometric mismatch is likely to constrain growth and reproduction of these littoral invertebrates. Therefore, the concept of stoichiometric food quality limitation might also be applicable to the littoral food web in lakes.

Key words: benthos, C:N:P ratio, ecological stoichiometry, herbivory, homeostasis, lake littoral, macroinvertebrates, mismatch, nutrient ratios, phosphorus

Introduction

Elemental nutrients, especially phosphorus (P) and nitrogen (N), are considered to be key elements in determining ecosystem productivity as the availability of phosphorus and nitrogen often limits primary production (ELSER et al. 1990). The primary producers have evolved mechanisms to cope with the fluctuating availability of elemental nutrients in their environment: They are able to store excess elemental nutrients (DROOP 1974) and improve their uptake (Fitzgerald and Nelson 1966). Therefore, the elemental composition of primary producers is variable, depending on environmental conditions (Elser et al. 2000b). In contrast, herbivores are often homeostatic, i.e., the C:P and C:N ratios vary over a narrow range, as has been shown for a variety of zooplankton taxa (Hessen 1990, Andersen and Hessen 1991). The requirement of the herbivores for elemental nutrients and the varying availability of these essential resources in their food can constrain their growth (Sterner and Schulz 1998) and reproduction (Færøvig and Hessen 2003). Such stoichiometric mismatches between primary producers with often highly variable elemental composition and homeostatic herbivorous consumers are considered to be a key factor that leads to poor energy transfer at the plant–herbivore interface in food webs (Sterner and Schulz 1998). STERNER & ELSER (2002) recently summarized the current knowledge on the interaction of mineral nutrients with molecules, organisms, and ecosystems. However, in contrast to the vast number of studies on zooplankton stoichiometry (e.g., Sterner and Elser 2002, and references therein), only little is known about the elemental composition of freshwater benthic invertebrates and their periphyton resource. In particular, the role of periphyton elemental nutrient stoichiometry as a food quality determinant for benthic freshwater invertebrates is not yet understood. To date, only two experimental studies have provided evidence for phosphorus limitation in mayflies (Frost and Elser 2002a) and in a lentic prosobranch (Stelzer and Lamberti 2002). However, no attempts have been made to determine the seasonal dynamics of the elemental nutrient stoichiometry of natural periphyton and associated invertebrates in a lake, which is a prerequisite to estimate the extent to which elemental nutrients might constrain food quality for herbivorous macroinvertebrates. To date, field studies have only analysed the elemental composition of either benthic invertebrates (FROST et al. 2003) or periphyton (KAHLERT 1998, KAHLERT et al. 2002), but have never considered both aspects simultaneously, which would allow hypotheses on possible stoichiometric limitations within the benthic food web to be formed. In a recent study, LIESS & HILLEBRAND (2005) reported nutrient ratios from littoral zone invertebrates and periphyton. However, they did not sample consumer and resource at the same dates and sites, which makes it difficult to relate the nutrient ratios of the periphyton to the elemental composition of the invertebrates. Our aim was to simultaneously study the elemental compositions of periphyton and invertebrate grazers over an entire growth season. Thus, for the first time, we directly related the stoichiometry of benthic herbivorous consumers to the stoichiometry of their periphyton resource to investigate, whether a mismatch between demand and supply of elemental nutrients occurs in the field. We present data from a field survey that was conducted during the growing season (April–October) in 2003 in the littoral zone of Lake Constance, a large lake in southern Germany. The aim of the study was to assess both temporal and spatial variation in the elemental nutrient stoichiometry of the dominant benthic, herbivorous invertebrates and simultaneously in their natural food resource, the periphyton community. We specifically addressed the hypotheses that i) the elemental composition of periphyton in the littoral zone of Lake Constance is variable both in space and time, ii) benthic herbivorous invertebrates maintain homeostatic body C:N:P ratios independent of the elemental composition of their periphyton resource, and, iii) there are species-specific differences in the elemental composition between groups of macroinvertebrate taxa.

Methods

Study sites

The study was conducted during the summer season (April–October) in 2003 in Upper and Lower Lake Constance, two basins of a large, pre-alpine meso-oligotrophic lake in central Europe. Upper Lake Constance is large and deep (surface area 473 km^2, mean depth 101 m), whereas Lower Lake Constance is smaller and shallower (surface area 63 km², mean depth 14 m). The site in the Lower Lake (N47°41.10', E9°04.19') is close to an agricultural drainage inflow, which led to thick periphyton layers on the surrounding cobblestones. The periphyton layer at the Upper Lake Constance site (N47°41.47', E9°12.20') was thinner and comprised higher amounts of inorganic sediments.

Sampling design

Both sites were sampled on the 15th day of each month from April through October 2003 at 40-cm water depth. Four replicate samples of periphyton and of invertebrates were taken and analysed, thus resulting in 2 sites × 7 months × 8 taxa × 4 replicate samples. Only groups of organisms were chosen that occurred frequently enough to allow reliable determination of body elemental nutrient ratios and that are considered to be at least partially herbivorous (MOOG 1995). The collected taxa were: *Asellus aquaticus* L. (Isopoda, Asellidae), *Bithynia tentaculata* L. (Prosobranchia, Bithyniidae), *Gammarus roeseli* Gervais (Amphipoda, Gammaridae), and *Tinodes waeneri* L. (Trichoptera, Psychomyidae). When a quick determination of live samples to the species level was not possible or when certain species occurred only at few sampling dates, groups of taxa were pooled at the level of orders or families. For example, all case-bearing Trichoptera, all Ephemeroptera, all Oligochaeta, and all Chironomidae were considered as groups dominated by herbivorous taxa in Lake Constance (BAUMGÄRTNER 2004) and analysed as groups. The sampled case-bearing Trichoptera were members of the Leptoceridae, Sericostomatidae, and Goeridae, the Ephemeroptera of the genera *Caenis* and *Baetis*. All groups were present at both sites, except for *A. aquaticus*, which occurred only in Lower Lake Constance.

Sample processing and analyses

Periphyton samples were taken with a syringe-brush sampler (PETERS et al. 2005), which scrapes a defined area (3.14 cm²) off natural hard substrates (cobblestones). Macroinvertebrate samples were taken non-quantitatively by collecting substrate from a 30×30 cm (900 cm²) area into a hand net. In addition, dissolved nutrients (soluble reactive phosphorus and nitrogen) in the water column approximately 10 cm above

ground were analysed. All samples were stored at 4° C in the dark until processing within 24 h of sampling.
To assess whether Lake Constance invertebrates that graze on periphyton could be limited by the quantity of their resource, periphyton standing stocks, i.e., chlorophyll *a* and ash-free dry mass, were determined as described by PETERS et al. (2005). The October samples for chlorophyll *a* and ash-free dry mass were lost during processing in the laboratory; therefore, biomass parameters are only available for the period from April through September.
Aliquots of the periphyton samples were filtered on precombusted glass-fibre filters (Whatman GF/F, 25 mm diameter, Whatman, Maidstone, UK) and dried for subsequent analysis of particulate organic carbon and particulate organic nitrogen with an NCS-2500 analyser (Carlo Erba Instruments). For determination of particulate phosphorus (P_{part}), aliquots were filtered through acid-rinsed polysulfone membrane filters (HT-200, Pall, Ann Arbor, Mich., USA) and digested with a solution of 10% potassium peroxodisulfate and 1.5% sodium hydroxide at 121 °C for 60 min, before soluble reactive phosphorus was determined using the molybdate-ascorbic acid method (GREENBERG et al. 1985).
Live macroinvertebrate samples were sorted. Snails and case-bearing caddisflies were removed from their shells and cases before analyses. Sorted invertebrates were stored frozen at –80 °C until analysis. Prior to analysis, samples were freeze-dried and ground to a powder. Aliquots of freeze-dried and powdered invertebrate bodies (approx. 1 mg each) were placed either into tin cups (HEKAtech, Wegberg, Germany) for C/N analysis or into glass vials for P_{part} determination as described above.

Statistical analyses

Periphyton nutrient ratios were ln(x) transformed to obtain homogeneity of variances and compared with a two-way ANOVA with date and site as fixed factors and the nutrient ratio as dependent variable, followed by a post-hoc comparison with Tukey's HSD (with α=0.05). To test whether elemental nutrient ratios of invertebrates were dependent on the elemental nutrient ratio of their periphyton resource, nonparametric correlations between the mean ratios for each sampling date and site were calculated, and the resulting Spearman rank correlation coefficients were tested for significance at α=0.05. The analyses of variance were performed using the GLM module of STATISTICA v.6 software package (STATISTICA, version 6, StatSoft, Inc. (2004), Tulsa, USA); Spearman's R in the correlations between periphyton and macroinvertebrate nutrient ratios was calculated with the nonparametric module of the same software.

An analysis of similarities (ANOSIM) test was applied to assess whether significant differences in the carbon, nitrogen and phosphorus composition of invertebrates between sites, sampling dates and species/groups were present. Similarity between samples was measured on matrices with relative (%) carbon, nitrogen and phosphorus content of invertebrates. Bray-Curtis coefficient S was used for computing similarity (with untransformed data), with a coefficient value of 100% for completely similar samples. The ANOSIM procedure compares the ranked similarities for differences within and between groups. The resulting R-value usually lies between 0 and 1, but can lie within a range of –1 to +1. An R-value of approximately 0 suggests an acceptance of the null hypothesis, whereas large R-values indicate separation of the groups, and small values close to 0 imply little or no separation (Clarke and Warwick 2001). Here, an R-value of > 0.6 was considered to be an indicator of pronounced difference between sample groups. In contrast to standard Z-type statistics, R has an absolute value interpretation of its value that is potentially more meaningful than its statistical significance. As with standard tests, R can be significantly different from 0 with a difference too small to be important, if there are enough replicates. There is a global R for the analyses according to the ANOVA result and for pairwise comparisons according to multiple post-hoc tests. This analysis was carried out using the PRIMER 5 software package (PRIMER version 5, PRIMER-E Ltd. (2001), Plymouth, United Kingdom).

Results

Periphyton biomass

Periphyton biomass at both sites was high in April, especially in Upper Lake Constance, and declined during May and June to 2.8 µg chlorophyll *a* cm^{-2} and 0.4 mg ash-free dry mass cm^{-2}, respectively (Fig. 1).

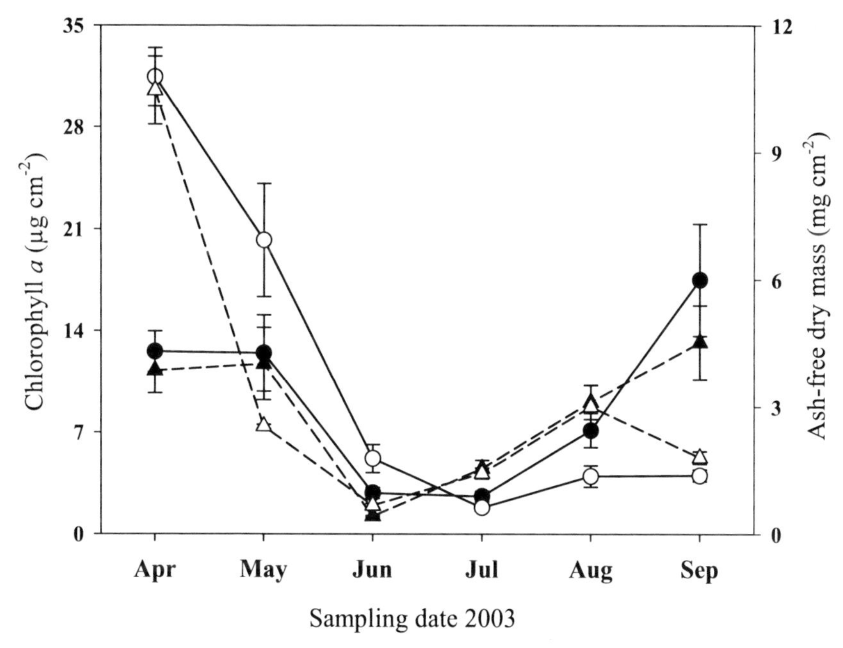

Figure 1: Biomass of Lake Constance periphyton measured as chlorophyll *a* (µg cm^{-2}, circles) and ash-free dry mass (mg cm^{-2}, triangles). Open symbols, samples from the Upper Lake site; filled symbols, samples from the Lower Lake site. Error bars represent ± 1 SE (of n=4 replicates).

During late summer (July–September), periphytic biomass recovered in the Lower Lake and finally reached values similar to those in spring. In the Upper Lake, the biomass remained low throughout the rest of the sampling period (Fig. 1). At this site, the periphyton biomass does not recover earlier than in November, when the high abundances of herbivorous invertebrates, especially gastropods, decline owing to low temperatures in the littoral zone (data not shown) and the decreasing water level in the lake (BAUMGÄRTNER 2004). The periphyton community was dominated by diatoms. Green algae occurred frequently, but were usually not dominant, and cyanobacteria were only of minor abundance.

Periphyton nutrient ratios

Not only the biomass of the periphyton, but also the elemental composition of the periphyton showed temporal and spatial variation (Fig. 2). Molar C:N and N:P ratios

Table 1: Results of two-way ANOVA on the mean molar elemental nutrient ratios of Lake Constance periphyton with sampling date and site as factors.

		C:P			C:N			N:P		
Factor	d.f.	MS	F	*p*	MS	F	*p*	MS	F	*p*
Date	6	0.649	16.83	< 0.001	0.102	4.19	< 0.01	0.720	14.039	< 0.001
Site	1	0.012	0.32	0.57	3.751	154.99	< 0.001	4.195	81.785	< 0.001
Date×Site	6	0.791	20.50	< 0.001	0.102	4.22	< 0.01	0.749	14.606	< 0.001
Error	37	0.039			0.024			0.051		

varied significantly between sites as well as C:P, C:N and N:P ratios between sampled months at the same sites as revealed by a two-way ANOVA (Table 1). C:P ratios were especially high at the Upper Lake site during spring and early summer (April–June), reaching means of up to 1225:1 in June (Fig. 2A).

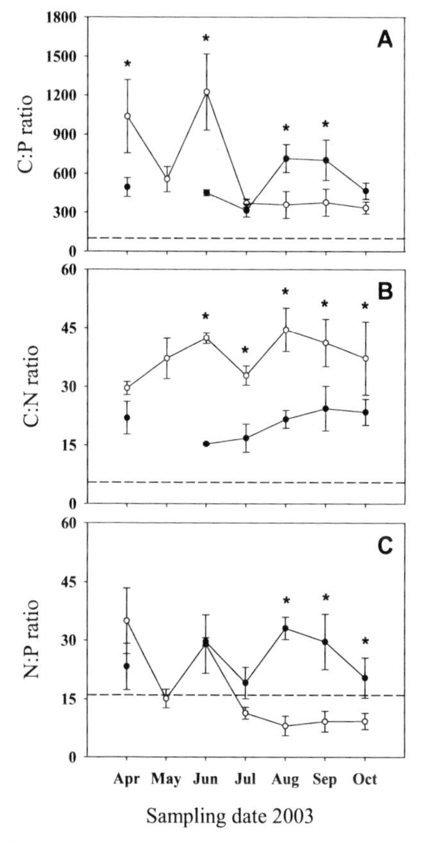

Figure 2: Molar elemental nutrient ratios (± 1 SE) of Lake Constance periphyton during the 2003 sampling period. Open circles, samples from the Upper Lake site; filled circles, samples from the Lower Lake site; dashed lines, Redfield ratio. A: C:P ratio; B: C:N ratio; C: N:P ratio. Asterisks indicate significant differences between sites (multiple comparisons among means with Tukey's HSD test after two-way ANOVA).

The values decreased during the summer months, and in late summer (August–September), the periphyton at the Lower Lake site had significantly higher C:P ratios than at the Upper Lake site (Fig. 2A). The C:N ratios of the periphyton from the Upper Lake site were significantly higher than those of the periphyton from the Lower Lake site from early summer until autumn (June–October, Fig. 2B). For all sites and dates, both C:P and C:N ratios were always well above the Redfield ratio (C:N:P = 106:16:1).The N:P ratio, which is considered as a measure to distinguish between nitrogen or phosphorus limitation in the periphyton (Hillebrand and Sommer 1999), showed a less-consistent pattern (Fig. 2C). During late summer (August–October), when the N:P ratios in the periphyton from the Upper Lake became very low (around 9:1), N:P ratios at the Lower Lake site were significantly higher than those in the Upper Lake. In general, N:P ratios of Lake Constance periphyton remained around the Redfield ratio of 16:1 (Fig. 2C). Despite the pronounced fluctuations in the nitrogen and phosphorus content of the periphyton, these changes seemed to be largely independent from the availability of dissolved reactive phosphorus and nitrogen in the water column, as dissolved nutrients constantly remained at low levels close to the detection limit throughout the whole summer. Average concentrations (± SE) of dissolved reactive phosphorus and nitrogen were 7.6 (± 3.2) and 742.1 (± 72.4) µg L^{-1}, respectively.

Invertebrate nutrient ratios

The variation in the elemental composition of the invertebrates was lower than that of the periphyton, especially for the groups where a determination to the species level was possible. When the standard deviation of the C:P ratio of the respective taxonomic group was normalized to the standard deviation of the C:P ratio of the periphyton, the variability in the invertebrate composition was only 8.2% (*A. aquaticus*), 6.2% (*G. roeseli*), and 9.6% (*T. waeneri*) of the variation found in the periphyton. In some cases, the variability was found to be higher for groups with a lower taxonomic resolution (Oligochaeta, 18.3%; and case-bearing Trichoptera, 15.0%). However, there was no clear connection between the observed variability in the invertebrates' C:P ratios and the taxonomic resolution, as the Chironomidae (9.1%) and the Ephemeroptera (4.2%) showed very low variability in C:P ratios despite the lower taxonomic resolution. The prosobranch snail *B. tentaculata,* with a comparatively high respective variability of 30.9%, was the only exception to this pattern.

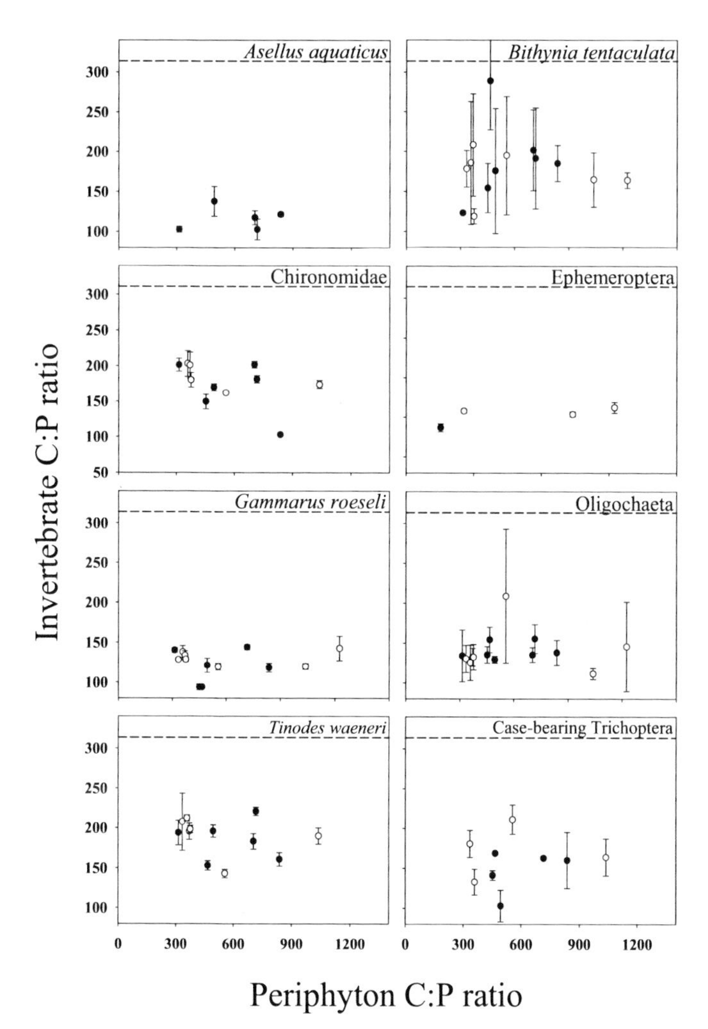

Figure 3: Molar C:P ratios (mean ± 1 SE) of Lake Constance herbivorous macroinvertebrates plotted against the molar C:P ratios of Lake Constance periphyton from the same samplings. Open circles, samples from the Upper Lake site; filled circles, samples from the Lower Lake site; dashed lines, lowest C:P ratio found in the periphyton samples (313.4).

The mean values of the C:P and C:N ratios of all analysed taxonomic groups were consistently lower than any of the C:P and C:N ratios found in the periphyton (Figs. 3,

4). The elemental composition of the periphyton was not reflected in the elemental composition of the invertebrates.

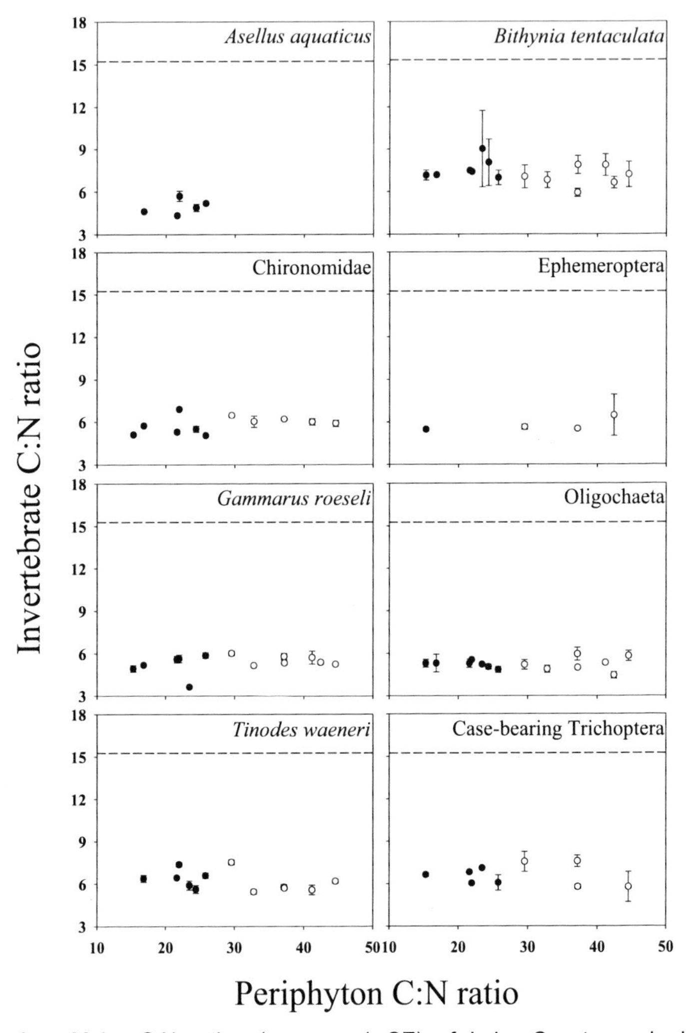

Figure 4: Molar C:N ratios (mean ± 1 SE) of Lake Constance herbivorous macroinvertebrates plotted against the molar C:N ratios of Lake Constance periphyton from the same samplings. Open circles, samples from the Upper Lake site, filled circles, samples from the Lower Lake site; dashed lines, lowest C:N ratio found in the periphyton samples (15.3).

Using a linear model, analyses between elemental nutrient ratios of periphyton and invertebrates revealed no significant correlations, neither for the C:P nor for the C:N ratios.

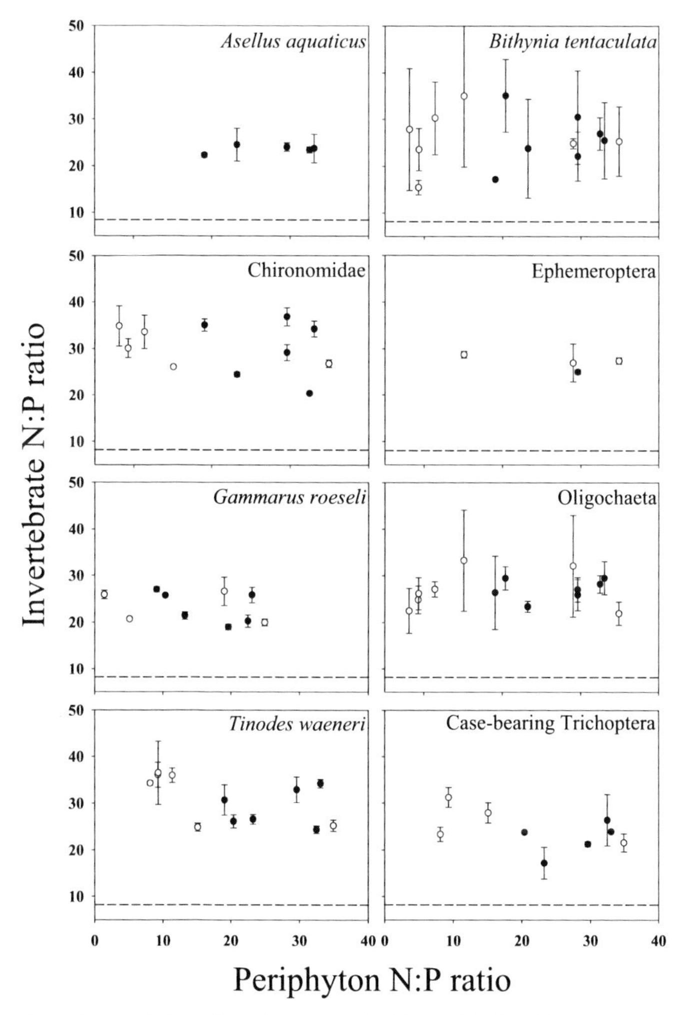

Figure 5: Molar N:P ratios (mean ± 1 SE) of Lake Constance herbivorous macroinvertebrates plotted against the molar N:P ratios of Lake Constance periphyton from the same samplings. Open circles, samples from the Upper Lake site; filled circles, samples from the Lower Lake site; dashed lines, lowest N:P ratio found in the periphyton samples (8.1).

Table 2: ANOSIM test (2-way crossed) for differences in carbon, nitrogen and phosphorus composition of invertebrates between sites (Upper and Lower Lake) and sampled taxa/groups (*Asellus aquaticus*, *Bithynia tentaculata*, Chironomidae, Ephemeroptera, *Gammarus roeseli*, Oligochaeta, *Tinodes waeneri* and Trichoptera). Pronounced differences ($R > 0.6$) are in bold. The R-value and the significance level (*, **, ***, with $p < 0.05, 0.01, 0.001$) are given for each combination.

Between Sites Upper vs. Lower Lake (Global R = 0.027*)

Betweeen Taxa/Groups (Global R = 0.38***)

	A. aquaticus	*B. tentaculata*	Chironomidae	Ephemeroptera	*G. roeseli*	Oligochaeta	*T. waeneri*
B. tentaculata	0.531 ***						
Chironomidae	0.060 n.s.	0.39 ***					
Ephemeroptera	**0.754** ***	**0.642** ***	0.067 n.s.				
G. roeseli	0.023 n.s.	**0.612** ***	0.279 ***	**0.647** ***			
Oligochaeta	0.108 n.s.	0.513 ***	0.086 **	-0.059 n.s.	0.175 ***		
T. waeneri	**0.906** ***	**0.815** ***	0.421 ***	0.289 **	**0.834** ***	0.383 ***	
Trichoptera	**0.846** ***	**0.679** ***	0.241 **	0.279 **	**0.734** ***	0.238 **	0.049 n.s.

For the N:P ratios (Fig. 5), only the correlation between the N:P ratio of the periphyton and the caseless trichopteran *T. waeneri* was significantly negative (R= –0.62, p<0.05). Nevertheless, there was no positive correlation between the elemental nutrient ratios of resource and consumer.

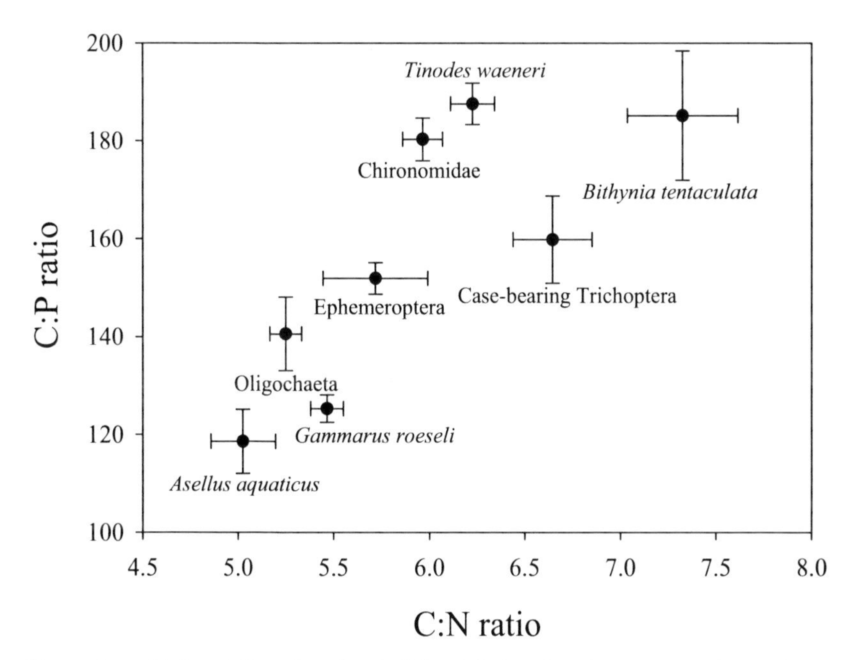

Figure 6: Molar C:N ratios (± 1 SE) of Lake Constance macroinvertebrates plotted against their molar C:P ratios. Data points represent means over all samplings ± 1 SE.

When the mean C:P ratio of the invertebrates was plotted against the mean C:N ratio, it became apparent that there were pronounced differences between the analysed taxonomic groups (Fig. 6). The observation of species-specific differences was supported by an analysis of similarity (ANOSIM) between the C, N and P contents (as % of dry weight) of the different invertebrate taxa and taxonomic groups (Table 2). This multivariate analysis revealed pronounced differences between some, but not all of the analysed taxa. However, two 1-way ANOSIMs (one for each site) did not indicate any effects of the sampling date (season) on the invertebrates' nutrient contents, as indicated by the low "Global *R*" values (Upper Lake Global R = 0.061 p<0.05; Lower Lake Global R = 0.015 n.s.).

Even though there were significant differences in the periphyton stoichiometry between the two sites (Table 1), no pronounced differences between the Upper and the Lower Lake site were found within the invertebrates (Table 2). Especially *G. roeseli*, *T. waeneri* and the Chironomidae showed almost constant body C:P and C:N ratios (Fig. 6). While the Ephemeroptera showed only little variation in their body C:P ratio, they seemed to be less constrained in their body C:N ratio (Fig. 6). In contrast to the mayflies, the Oligochaeta seemed to be more constrained in their body C:N ratio than in their C:P ratio (Fig. 6). Some of the invertebrates, such as the prosobranch snail *B. tentaculata,* exhibited a broader range in their body elemental nutrient ratios than those mentioned above, but even *B. tentaculata* was still a rather homeostatic consumer compared to the high variation in its periphyton food resource. In concordance with the observed differences in the C:P and C:N ratios of the invertebrates (Table 2, Fig. 6), the carbon, nitrogen, and phosphorus contents expressed as the percent of the body dry weight were also quite different (Table 3). In particular, *B. tentaculata* had lower carbon, nitrogen, and phosphorus contents (in its soft body) relative to the dry mass than any other taxon or group.

Discussion

Periphyton biomass

The biomass of Lake Constance periphyton showed considerable temporal and spatial variability. Therefore, it is important to disentangle possible effects of food quantity and food quality on the grazers. However, this is difficult, as the quantitative food requirements of many benthic invertebrates are not known and the functional responses of their food uptake have not been experimentally determined. In the littoral zone of Lake Constance, the periphyton biomass never was lower than 0.4 mg ash-free dry matter cm^{-2} (Fig. 1). We can use this for some indirect estimate using evidence from the literature. For example, when gastropods, which graze on periphyton very efficiently, were offered limiting quantities of periphyton, only 0.1 mg ash-free dry matter cm^{-2} remained (Stelzer and Lamberti 2002), which suggested that the 0.4 mg ash-free dry matter cm^{-2} found in this study indicates that food quantity was not limiting. Furthermore, the observed periphyton biomass of 0.4 mg ash-free dry matter cm^{-2} is similar to the food quantities (0.15 mg C cm^{-2}) that proved to be sufficient to detect significant food quality effects on the growth of larvae of the mayfly *Caenis* when fed algae with a high and a low C:P ratio (Frost and Elser

2002a). Hence, it can be assumed that during the entire sampling season, food quantity was sufficiently high in Lake Constance to rule out limitation of invertebrate growth by low food quantity.

Table 3: Carbon, nitrogen, and phosphorus content of the bodies of herbivorous macroinvertebrates in Lake Constance. Values are given as means (± SE) of weight normalized to dry mass of animals pooled over all samplings.

Taxonomic group		% C	% N	% P
Asellus aquaticus	(n=14)	39.7 ± 0.7	9.3 ± 0.2	0.89 ± 0.03
Bithynia tentaculata	(n=48)	26.4 ± 1.1	4.6 ± 0.3	0.43 ± 0.02
Chironomidae	(n=38)	39.7 ± 1.6	7.8 ± 0.3	0.59 ± 0.03
Ephemeroptera	(n=15)	47.1 ± 1.1	9.9 ± 0.4	0.81 ± 0.03
Gammarus roeseli	(n=44)	39.3 ± 0.4	8.5 ± 0.1	0.83 ± 0.02
Oligochaeta	(n=52)	40.7 ± 1.1	9.2 ± 0.3	0.79 ± 0.03
Tinodes waeneri	(n=45)	51.1 ± 0.5	9.7 ± 0.2	0.72 ± 0.01
Case-bearing	(n=25)	49.6 ± 1.4	8.9 ± 0.3	0.85 ± 0.04

Periphyton elemental nutrient ratios

Not only the biomass, but also the elemental nutrient ratios of the periphyton showed high temporal and spatial variation (Fig. 2), thus supporting our first hypothesis. The mean C:P ratio of periphyton at the Upper Lake site reached 1225:1 during June 2003, which is even above the maximum values that FROST & ELSER (2002b) found in an ultraoligotrophic lake and much higher than values reported in the review of KAHLERT (1998) on freshwater periphyton nutrient ratios. Compared to the optimal C:N:P ratio for benthic algae of 119:17:1 (Hillebrand and Sommer 1999), the mean ratio of Lake Constance periphyton of 589:22:1 suggests a nutrient limitation of the autotrophic community. However, compared to carbon, both nitrogen and phosphorus contents were low; a distinction between nitrogen or phosphorus limitation according to HILLEBRAND & SOMMER (1999) is difficult. The photoautotrophs in the periphyton might have been co-limited by nitrogen and phosphorus. It should be noted that periphyton samples taken with the brush sampler can contain varying proportions of living components, such as photoautotrophs (algae and cyanobacteria), meiofauna organisms (rotifers, nematodes, etc.), and dead organic matter (detritus) as already discussed by KAHLERT (1998). A variable proportion of detritus versus autotrophic components might have further contributed the extremely high variation in periphyton nutrient ratios. However, for the scraper and collector taxa studied here, the entire periphyton community sampled by the brush sampler is probably similar to what the invertebrates collected as their food, provided that the invertebrates did not actively select for nutrient- rich particles.

Invertebrates

The elemental nutrient ratios of the invertebrate bodies varied only within very small ranges, which was in remarkable contrast to the highly variable elemental nutrient ratios found in the samples of their periphyton food source from the same sites. The only study to date in which the elemental nutrient stoichiometry of the body of an herbivorous invertebrate and its periphyton food were analysed simultaneously revealed a similar discrepancy: FROST et al. (2002a) found caddisfly larvae to have a body C:P ratio of 171:1 at a mean C:P ratio of the epilithic community of 820:1. Here, we present data that extend the findings of FROST et al. (2002a) to a broad range of benthic metazoan grazers. Admittedly, direct comparisons of resource and consumer stoichiometry might not equally apply to all taxonomic groups, considering possible species-specific bioenergetics. However, as there is no data available on metabolic rates of most benthic invertebrates, we found this straightforward approach of direct comparisons justified. The small variation observed in all the analysed groups of invertebrates indicates that these organisms maintain stoichiometric homeostasis of their body tissue. We can not exclude the possibility that the invertebrates feed selectively on resources with a constant elemental composition, and therefore reflect this (constant) elemental composition in their body tissue without actively maintaining homeostasis. However, this seems unlikely, as all the analysed taxonomic groups belong to the functional groups of either scrapers, brushers, and/or shredders and are all considered to be rather unselectively ingesting the whole periphyton community (e.g., BARNESE et al. 1990, DIAZ VILLANUEVA et al. 2004).
The analysed taxonomic groups showed differences in the degree to which their elemental nutrient ratios varied, and, in contrast to the findings of HESSEN & LYCHE (1991) for freshwater zooplankton, not only the C:P ratios, but also the C:N ratios of the benthic invertebrates differed between the taxonomic groups. These differences in (homeostatically fixed) body nitrogen and phosphorus content probably reflect

species-specific physiological requirements of the different taxonomic invertebrate groups. These different requirements are likely to influence their susceptibility to competitive exclusion under elemental nutrient limitation, as it has been found for freshwater zooplankton (STERNER 1990). This might be especially interesting since, in contrast to zooplankton (Hessen and Lyche 1991), apparently not only the phosphorus content, but also the nitrogen content (relative to carbon) differs between investigated taxa, which suggests that competition for the two nutrients might have important consequences for the community structure of benthic herbivores. In particular, further investigations might reveal if the growth of certain invertebrate taxa feeding on the more nitrogen-rich periphyton at the Lower Lake site is limited by the availability of phosphorus alone, while the invertebrates might be co-limited by both nutrients at the Upper Lake site, where both phosphorus and nitrogen are equally low in the periphyton. Different requirements for phosphorus and nitrogen might also contribute to the observed changes in the composition of the Lake Constance macroinvertebrate community in both season and depth (BAUMGÄRTNER 2004). For example, the relatively high abundance of the caseless caddisfly *Tinodes waeneri* compared to the case-bearing Trichoptera (BAUMGÄRTNER 2004) might be partly due to their observed lower P requirements (Fig. 6). However, this remains to be resolved by further experimental studies on the competitive abilities of these taxa under different nutrient availabilities. The higher variation in the elemental nutrient ratios in the prosobranch gastropod *B. tentaculata* than in the other groups of invertebrates might be explained by the comparatively long life expectancy of several years of *B. tentaculata* (Brendelberger 1997a), and therefore, probably several age-classes of this taxon were pooled in the analyses. Probably, a sorting of *B. tentaculata* size classes would have reduced this variation. For other taxa (e.g., copepods, CARRILLO et al. 1996), ontogenetic shifts in the elemental composition of the body occur with the age of the individual. This might have also been the case for *B. tentaculata*, thus leading to a larger variation in the elemental nutrient ratios of the body tissues. Furthermore, a relatively high variability in body stoichiometry relative to other taxa might be a competitive advantage, if high C:N:P ratios in the periphyton resource are common. The fact that *B. tentaculata* was the least homeostatic taxon in our study might therefore, at least in part, explain why it is one of the most abundant benthivorous invertebrates in Lake Constance (BAUMGÄRTNER 2004).

The choice of macroinvertebrate groups for this study was constrained by several factors. Four taxonomic groups were not sorted to the species level since either only single specimens from different species were found at most of the sampling dates (Trichoptera, Ephemeroptera) or a determination of live samples to the species level was not possible (Oligochaeta, Chironomidae). No Ephemeroptera were found later than June, which can be explained by emergence. Emergence during the sampling period also occurred for *T. waeneri* (June), the case-bearing Trichoptera, and parts of the chironomid population. Therefore, for some of the insect larvae, it was not possible to obtain samples of these taxa in all the sampled months.
The analysed invertebrates can all be considered as at least partially herbivorous (MOOG 1995), i.e., they rely on periphyton as food source. Furthermore, there were no leaves present as an alternative food source for shredders, such as *G. roeseli* and *A. aquaticus,* at either of the two sites during the sampling period. Therefore, it can be assumed that the invertebrates had to rely largely on periphyton as the food source. Detritus, which can make up a large proportion of periphyton assemblages, is known to be even more depleted in P and N than benthic algae (KAHLERT 1998), so that this study might even have underestimated the effective stoichiometric mismatch for selectively detritivorous taxa. Other macroinvertebrates, e.g., the zebra mussel (*Dreissena polymorpha*), leeches (Hirudinea), and tricladids (Turbellaria), occur frequently in Lake Constance (BAUMGÄRTNER 2004), but these were not analysed since they are either predominantly carnivorous (MOOG 1995) or active filter-feeders of phytoplankton, and therefore, do not fit in the context of a possible coupling of periphyton and benthic grazer stoichiometry.

Possible consequences of stoichiometric mismatch

Mismatches between resource and consumer nutrient stoichiometry will likely result in strong effects on growth and reproduction of the invertebrate consumers, which in turn will influence important ecosystem processes such as nutrient turnover. Such consequences are especially likely to occur when species with high somatic growth rates are affected. The "growth rate hypothesis" states that species with higher specific growth rates need a higher percentage of P-rich RNA in their body tissue than slower-growing species (ELSER et al. 1996). Therefore, their demand for phosphorus in the diet is higher (a lower C:P ratio in the food is needed), and low food quality in terms of high food C:P ratios will affect the growth of those fast-growing species more than the growth of species with a lower growth rate. For the species-specific growth rates of the taxonomic groups analysed here, there is no data available in the literature except for *B. tentaculata*, with maximum somatic growth rates of 0.033 d^{-1} (Fink and Von Elert *in press*) and the mayfly *Caenis* with somatic growth rates of up to 0.04 d^{-1} (Frost and Elser 2002a). In the future, it will be interesting to test (*i*) whether the growth rate hypothesis can be applied to macroinvertebrates, i.e., whether species with a high phosphorus content grow faster than species with a low phosphorus content, (*ii*) whether fast-growing benthic invertebrates are more susceptible to phosphorus limitation than taxa with lower growth rates, and (*iii*) whether the correlation between body tissue phosphorus content and growth rate can be broken by nitrogen limitation, as has been suggested by ACHARYA et al. (2004).

A dietary mismatch between grazers and periphyton not only affects growth and reproduction of the grazers, but also affects grazer-mediated nutrient regeneration and thus has whole-ecosystem effects. Several studies have investigated this interaction (e.g., Hillebrand and Kahlert 2001, Frost et al. 2002a, Hillebrand et al. 2004). However, until now, the ecological importance of these effects was unclear owing to the lack of field data on the elemental nutrient stoichiometry of both herbivorous invertebrates and periphyton. The results of this study allow grazer-mediated nutrient regeneration to be related to species-specific differences in invertebrate elemental nutrient content and to periphyton elemental nutrient ratios actually occurring in the field.

It should be noted that the magnitude of effects of stoichiometric mismatch on growth and reproduction of consumers and on nutrient recycling would be affected by compensatory feeding behaviour of the invertebrates, as has been found for marine amphipods (Cruz-Rivera and Hay 2000), daphnids (Plath and Boersma 2001) and freshwater gastropods (FINK & VON ELERT unpubl.). By ingesting more of a low-quality food (e.g., periphyton with a high C:P ratio), consumers could still satisfy their need for limiting essential nutrients, thus dampening the effects of dietary mismatch on growth and reproduction. On the other hand, an increase in grazing activity might enhance recycling of nutrients that are not limiting for the grazer and could thereby lead to shifts in the kind of nutrient limitation in other members of the benthic invertebrate community. Hence, a mismatch between the elemental composition of primary producers and herbivores can hamper the energy transfer at the plant–

herbivore interface, influence competition between grazers, and alter the regeneration of limiting nutrients. It is high time that these concepts, extensively studied in freshwater plankton, are also tested for their applicability in benthic systems. The confirmation of grazer homeostasis and of highly variable elemental nutrient ratios of the periphytic resource by the data presented in this study allows further testing of hypotheses on the role of food quality limitation of herbivore growth and reproduction, grazing-mediated nutrient recycling, and whole-ecosystem effects of changing nutrient availability in freshwater benthic ecosystems owing to eutrophication (nitrogen, phosphorus) or climatic change (carbon).

Acknowledgements

We thank C. Gebauer and O. Walenciak for excellent technical assistance, K. Brune for editing the English, and N. Scheifhacken and M. Korn for help with the identification of invertebrate taxa. Two anonymous reviewers gave helpful comments that improved the manuscript considerably. This study was supported by the German Research Foundation (DFG) within the Collaborative Research Centre SFB 454 “Littoral Zone of Lake Constance”.

Chapter 3

Stoichiometric constraints in benthic food webs: nutrient limitation and compensatory feeding in the freshwater snail *Radix ovata*

Patrick Fink and Eric von Elert

Abstract

Nitrogen (N) and phosphorus (P) are considered to be essential nutrients that control secondary production in freshwater plankton; insufficient availability of N and P can limit herbivore growth. However, to what extent the dietary stoichiometry determines the food quality for benthic consumers is not clear. Here, data are presented from field samplings and from a laboratory experiment on the potential of primary producers low in P, N, or P and N to constrain growth of the freshwater gastropod *Radix ovata*. The filamentous green alga *Ulothrix fimbriata* was cultured under different nutrient regimes, resulting in algae with different C:N:P ratios. The pure algae were fed in high and low quantities to juvenile *R. ovata*. Low availability of N and especially P in the algae strongly constrained the biomass accrual of the herbivore. In accordance with theoretical predictions, these food quality differences were highly dependent on the food quantity. The snails' growth rate was significantly related to their body C:P ratio, thereby supporting the Growth Rate Hypothesis. *R. ovata* displayed a pronounced compensatory feeding response to low-nutrient food that could partly dampen but not fully compensate the food quality effects on snail growth. Increased feeding of gastropods at low P and/or N availability leads to depletion of periphyton biomass; hence compensatory feeding would shift the benthic herbivore community from a P or N limitation to a C limitation and thus have whole-ecosystem effects.

Introduction

To understand the energy transfer at the plant–herbivore interface in food webs, it is essential to determine the factors that control the efficiency of the energy transfer. The energy transfer efficiency is highly variable, often because of the nutritional quality of the food for the consumers. The nutritional quality can be measured as the resulting growth or reproductive output per unit of consumed resource e.g., carbon (C). However, effects of food quality and food quantity on herbivore fitness are often difficult to separate (Sterner 1997, Sterner and Schulz 1998).

The factors determining food quality of freshwater zooplankton have been extensively studied (Sterner and Schulz 1998). The mineral nutrients nitrogen (N) and phosphorus (P) have gained attention because their limited availability leads to highly variable C:N:P ratios in primary producers. High C:N and C:P ratios in primary producers in a broad range of ecosystems (terrestrial insect–herbivore interactions, and marine and especially freshwater plankton) could result in insufficient availability of N and P, which can limit herbivore growth (Sterner 1993, Elser et al. 2000b).

However, although littoral algal communities are among the most productive assemblages in aquatic ecosystems (Pinckney and Zingmark 1993), to what extent the dietary stoichiometry determines the food quality for benthic consumers is not clear. In order to gain a deeper understanding in the structure of the highly dynamic and productive littoral zone habitats (Wetzel 2001), it is crucial to understand how the trophic transfer of littoral primary production through the benthic food web is regulated, and which factors are responsible for constraints in this energy transfer. Thus, in order to gain a deeper understanding of algae-herbivore interactions in this benthic ecosystems, further investigations on the biological stoichiometry (sensu Elser et al. 2000a) of macroinvertebrates are necessary (Frost et al. 2002b).

Several mechanisms for overcoming growth limitation of consumers caused by low nutritional quality have been suggested (e.g., Sterner and Hessen 1994). Herbivores could actively select particles with high contents of the limiting mineral nutrient (Butler et al. 1989) or enhance nutrient retention during digestion and thus release feces that are nutrient-depleted relative to the food source. Another strategy could be an

increased uptake of food with low nutritional quality to access sufficient amounts of the limiting nutrient. Even though such active compensatory feeding has the potential to shift the whole community from an N or P limitation into a C or energy limitation and thus, probably whole-ecosystem consequences, this mechanism has not yet been explicitly demonstrated in the context of biological stoichiometry.
A major assertion of the Ecological Stoichiometry Theory (Elser et al. 2000a) is the Growth Rate Hypothesis, which states that organisms with a higher specific growth rate have a high content of RNA with a P content higher than in slower-growing organisms. This concept is supported by data from a variety of organisms ranging from microbes to insects (Elser et al. 2003). Such effects of specific growth rates on somatic C:P ratios are not restricted to interspecific differences because also intraspecific changes in P content occur with different growth rates (DeMott et al. 1998). Evidence for the validity of the Growth Rate Hypothesis mainly comes from studies of terrestrial systems and freshwater zooplankton (Elser et al. 2003), and the hypothesis has not been experimentally tested with freshwater benthic organisms. Here, data are presented from field samplings and from a laboratory experiment on the potential of a primary producer (the filamentous green alga *Ulothrix fimbriata)* low in N, P, or both nutrients to constrain the growth of a freshwater benthic herbivore, the pulmonate gastropod *Radix ovata* (DRAPARNAUD), and to change the herbivore's nutrient assimilation efficiency. *Radix* is a dominant genus of benthic grazers that feed on periphyton (Calow 1970) in many lakes and rivers of temperate Europe and has widely spread in North America after being introduced during the 19th century. Furthermore, we investigated whether *R. ovata* can compensate for low nutritional quality by ingesting more of a low-quality food to obtain sufficient amounts of limiting nutrients. Specifically, we hypothesized that i) low availability of P and/or N would constrain the biomass accrual of *R. ovata*, ii) the body C:N:P stoichiometry of *R. ovata* would be correlated with the snail's growth rates and not with the C:N:P stoichiometry of the algae, and iii) the snails would respond to low dietary nutrients by excretion of nutrient-depleted fecal pellets and by increased food consumption to alleviate the effects of low P and/or N availability.

Methods

Field sampling

The primarily algivorous (Calow 1970) gastropod *Radix ovata* (DRAPARNAUD) was sampled over an entire growing season (April–October) in 2003 in the littoral zone (sampling depth approx. 40 cm) of Lake Constance, a large, pre-alpine, meso-oligotrophic lake in central Europe, to determine seasonal changes in the C:N:P ratios of the somatic tissue. The C:N:P ratios were compared to those of periphyton sampled from the same site on the same dates. Four independent replicate samples of both snails and periphyton were taken once per month with a periphyton brush sampler and by collecting snails from littoral hard substrates as described in detail by Fink et al. (*in press*). All snails within one replicate sample were pooled for the analyses. For the laboratory experiment, juvenile *R. ovata* (shell length 6.6 – 7.4 mm) were collected at the same site and depth. Juvenile snails have higher specific growth rates than adults (Brendelberger 1995b) because they invest most of their energy in growth and not in reproduction; they therefore have a higher specific P content, which may increase their susceptibility to P limitation compared to adults (Sterner and Schulz 1998).

Food cultures

The filamentous green alga *Ulothrix fimbriata* (BOLD) was obtained from the Göttingen (Germany) Algal Culture Collection (SAG 36.86) and cultivated semicontinuously (dilution rate 0.25 d^{-1}) on WC medium (Guillard and Lorenzen 1972) at 20°C with a light intensity of $1 \cdot 10^{16}$ quanta s^{-1} cm^{-2}. This species was chosen, first because *Ulothrix* is a common genus of filamentous, benthic green algae, occurring frequently in periphyton communities of lake littoral zones and second, it was known to support high growth rates for juvenile *R. ovata* (P. Fink, unpubl. results). Low-nutrient algae were obtained by culturing in P-free WC medium (K_2HPO_4 replaced by 0.04 mM KCl), N-free WC medium ($NaNO_3$ replaced by 1 mM NaCl), or P- and N-free WC medium (N and P replaced as above). Depleted WC media (2 L) was inoculated with a small amount of non-limited algae and incubated for one week until the cultures entered the stationary growth phase . Cells were harvested by centrifugation at 4000×g and resuspended in 0.45-µm filtered Lake Constance water (in which concentrations of dissolved P and N were around the detection limit during the summer months when the study was done); carbon concentrations were estimated from photometric light extinction at 800 nm and from carbon-extinction equations determined previously.

Growth experiments

Within 24 h of sampling, the shell length of the snails was measured to the nearest 0.01 mm using a dissecting microscope equipped with an image analysis system. The shell length is defined as the distance from the apex to the most distal part of the shell's aperture. Soft bodies of 32 individuals in the same size range as the snails used in the growth experiments were removed from their shells, freeze dried, and stored at –80°C until further analyses to determine initial values.
During the experiments, the snails were kept individually in 400-ml polyethylene containers (Buchsteiner, Gingen, Germany) containing 200 ml of 0.45-µm filtered Lake Constance water. The containers did not release any P or N into the water as tested in a preliminary experiment: after 48 hours incubation time in the containers, no dissolved P and N could be detected in 200 ml of ultrapure water with the molybdate-ascorbic acid method (Greenberg et al. 1985). Freshly prepared food algae were added daily at a biomass equivalent to 2 mg particulate organic carbon (POC) for the high quantity treatments and 0.5 mg POC for the low quantity treatments. The containers were kept at 20 ± 0.5°C under dim light to minimize algal growth. The experiment consisted of eight treatments (P-, N-, or P- and N-limited and nonlimited, with either high or low food quantity) with eight replicates each, i.e., 64 containers, each containing one *R. ovata* individual.
The water, feces, and remaining food were replaced daily with freshly filtered Lake Constance water and new food. The snails were transferred to new containers twice a week. The experiment was terminated after 35 days when the fastest growing animals reached the maturation size. The final shell length was measured as described above and soft bodies were then removed from the shells under a dissecting microscope and frozen immediately at –80°C. Frozen soft bodies were freeze-dried; the dry weight was determined with a microbalance (Mettler UTM2) to the nearest microgram. Growth of the juvenile snails was determined as the incremental increase in shell length (mm d^{-1}) and as somatic growth rate (d^{-1}).

The juvenile somatic growth rate (*g*) was determined as the increase in dry weight (W) from the beginning of the experiment (W_0) to day 35 (W_t) using the equation:

$$g = \frac{\ln W_t - \ln W_0}{t}.$$

Similar to studies with zooplankton, *g* can probably be used as a good estimate of fitness (Lampert and Trubetskova 1996), since the growth of juvenile *Radix* sp. follows an exponential function similar to *Daphnia* sp. (Brendelberger 1997a). Both parameters were determined as a snail's soft body mass is not directly related to its shell length, as soft body mass may vary depending on the feeding conditions (Brendelberger 1997a). Shell lengths and soft body dry weights of snails remaining from the initial field sampling were used to establish a linear log/log regression of shell length vs. soft body dry weight ($r^2 = 0.95$). Because the initial soft body dry weights (W_0) of the snails in the growth experiment could not be determined directly, W_0 was calculated using the regression and the shell lengths measured at the start of the experiment. After dry weight determination , the soft bodies were ground to a powder for elemental analyses.

Food consumption

At the beginning of the experiment, the snails were so small that even a small amount of food would not have been limiting and stored mineral nutrients would have buffered any nutrient deficiency in the food algae. Therefore, the amount of food consumed by *R. ovata* was measured close to the end of the experiment on day 33. Food consumption was determined as the difference between the amount of food offered and the food remaining after 24-h in the snails' containers, normalized to the biomass of the snails, resulting in biomass-specific consumption rates (mg C mg dry wt.$^{-1}$ d^{-1}). Snail body dry weight at day 33 (when food consumption was determined) was estimated by subtraction of daily growth increments from the final dry weight determined on day 35 of the growth experiment.

Elemental analyses

C:N:P ratios of *U. fimbriata* were analyzed as described by Wacker and von Elert (2001). C:N:P ratios of *R. ovata* soft bodies and feces were determined from ground freeze-dried samples . Approximately 1 mg of ground sample was placed into tin cups (HEKAtech, Wegberg, Germany) for C/N analysis and into glass vials for determination of particulate P (Wacker and von Elert 2001).

Statistical analyses

C:N:P ratios of field-collected and laboratory-grown *R. ovata*, of the food algae, and of snail fecal pellets were ln(x) transformed to obtain homogeneity of variances and compared using a one-way ANOVA, followed by a post-hoc comparison with Tukey's HSD. *R. ovata* shell ($\sqrt{(x)}$ transformed) and somatic growth rates were analyzed with a two-way ANOVA with food quantity and food nutrient regime as factors and the growth rate as the dependent variable, followed by a post-hoc comparison with Tukey's HSD. Biomass-specific rates of food consumption were analyzed in the same way without prior transformation of data.

All statistical analyses were performed using the GLM module of STATISTICA v.6 software package (STATISTICA, version 6, StatSoft, Inc. (2004), Tulsa, USA) and a significance level of α=0.05. The relationship between growth rate and body C:P ratio of the experimental animals was investigated using a linear model. Spearman's Rank Correlation Coefficient was calculated and tested for significance at α=0.05 using the module "Nonparametric Statistics" of the same software package.

Results

C:N:P ratios of R. ovata in the field

The C:N:P ratios of *R. ovata* soft bodies and of periphyton from the same samplings were compared. The C:P and C:N ratios of the periphyton showed a pronounced seasonal variation (Fink et al. *in press*), whereas the C:N:P ratios of *R. ovata* soft bodies remained constant independent of their food source's C:N:P ratios (Fig. 1A, B), which indicates strongly maintained homeostatic C:P and C:N ratios of *R. ovata* body tissue. Furthermore, *R. ovata* C:P ratios (Fig. 1A, $F_{(1, 39)}$=216.6, p<0.001) and C:N ratios (Fig. 1B, $F_{(1, 39)}$=3510.6, p<0.001) were lower than the respective nutrient ratios of the periphyton (Fig. 1A, B).

C:N:P ratios of U. fimbriata

The C:N:P ratios of the food algae were measured on day 6, 25, and 28 of the growth experiment. The C:P ratios of the algae grown in medium lacking P were higher (Tab.

1) than those of the algae grown in medium containing P. Algae grown either in the presence or in the absence of both N and P had similar N:P ratios (Tab. 1). The N:P ratio of algae grown in medium lacking P was higher than in the other cultures. The N:P ratio of algae grown in medium lacking N was lower than in the other cultures (Tab. 1). Nutrient depletion did not influence the morphology of *U. fimbriata* filaments as observed by light microscopy.

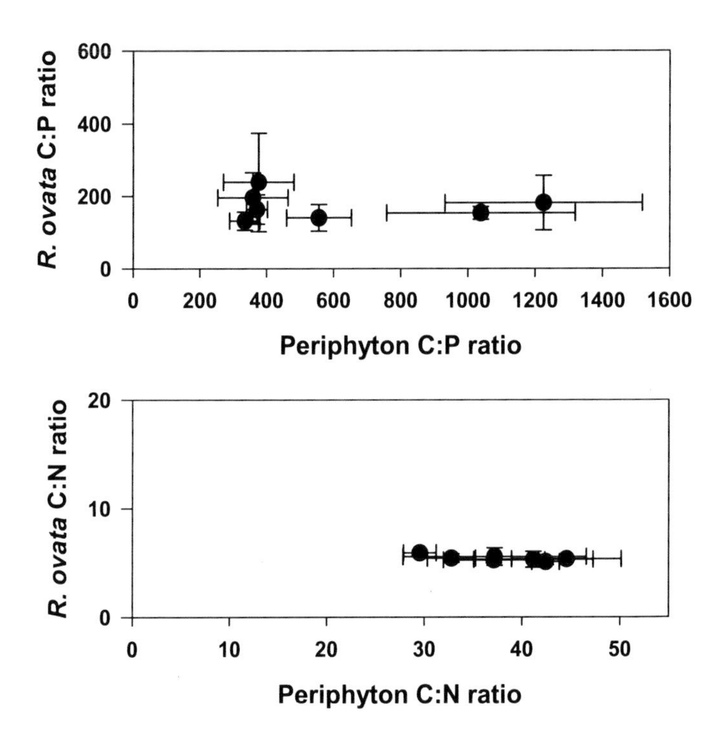

Figure 1: Molar carbon:nutrient ratios (mean ± SD; n=4 samples from the same date and site) of field-collected *R. ovata* versus the molar carbon:nutrient ratios of Lake Constance periphyton (mean ± SD of n=4) from the same sampling date and site; (A) C:P ratios; (B) C:N ratios.

Snail growth

The shells of juvenile *R. ovata* fed with high quantities of non-nutrient-depleted *U. fimbriata* grew faster than in any other treatment (Fig. 2A, Tab. 2). However, even low quantities of non-nutrient-depleted food supported higher growth than any of the nutrient-depleted food treatments, irrespective of food quantity. The growth rates of snails fed high quantities of algae grown in the absence of P and/or N did not differ from the respective growth rates of snails fed low quantities (Fig. 2A). N-depleted algae had less severe effects on *R. ovata* shell growth than P-depleted algae (Fig. 2A). The lowest snail growth occurred on P-depleted algae. Surprisingly, snails fed algae low in both N and P had higher shell growth rates than snails fed *U. fimbriata* grown without P, but with N in the culture medium (Fig. 2A).

Table 1: Mean molar C:N:P ratios of *U. fimbriata* fed to *R. ovata* in the four nutrient treatments. Values given are means (± SD) of day 6, 25 and 28 of the experiment. Letters indicate homogenous groups from Tukey's HSD test following significant effects (C:P $F_{(3, 8)}$=50.7, p<0.001, C:N $F_{(3, 8)}$=16.9, p<0.001, N:P $F_{(3, 8)}$=185.0, p<0.001) of a one-way ANOVA on ln(x)-transformed algal nutrient ratios.

Nutrient treatment of the alga	Mean nutrient ratio			Homogenous groups		
	C:P (± SD)	C:N (± SD)	N:P (± SD)	C:P	C:N	N:P
+P +N	245 (27)	13 (3)	20 (2)	A	A	A
+P –N	197 (54)	35 (7)	6 (1)	A	B	B
–P +N	1243 (414)	26 (9)	48 (3)	B	B	C
–P –N	942 (89)	42 (3)	22 (3)	B	B	A

The somatic growth rates of snails fed high quantities of non-nutrient-depleted non-nutrient-depleted *U. fimbriata* were higher than in any other treatment (Fig. 2B, Tab. 2). Somatic growth rates of snails fed low quantities of the same food suspension (+P+N) were lower than when fed high quantities, which indicated that growth in the low quantity +P+N treatment was limited by food quantity. Growth rates were comparably lower in all treatments where *R. ovata* was fed nutrient-depleted algae. Similar to the observations of the shell growth rate, food quantity did not affect the soft body growth rate when snails were fed nutrient-depleted algae (Fig. 2B).

Snail soft body C:N:P ratios

Food quantity did not affect the soft body nutrient stoichiometry in growth experiments (one-way ANOVA, C:P: $F_{(1, 54)}$=1.7, p=0.202; C:N: $F_{(1, 54)}$=0.1, p=0.748). Hence, data from the treatments with high and low quantities of algae of each algal nutrient regime were pooled and compared via one-way ANOVA. The C:P and C:N ratios of the soft bodies of the experimental snails were significantly affected by the nutrient regime of the food algae (C:P: $F_{(5, 95)}$=26.0, p<0.001; C:N: $F_{(5, 95)}$=134.2, p<0.001). Post-hoc tests revealed that C:P and C:N ratios increased when snails were fed nutrient-depleted algae for 35 days (Fig. 3A, B). However, the C:N and C:P ratios of the soft bodies generally increased when snails were fed nutrient-depleted algae irrespective of the actually omitted nutrient in the algal growth medium.

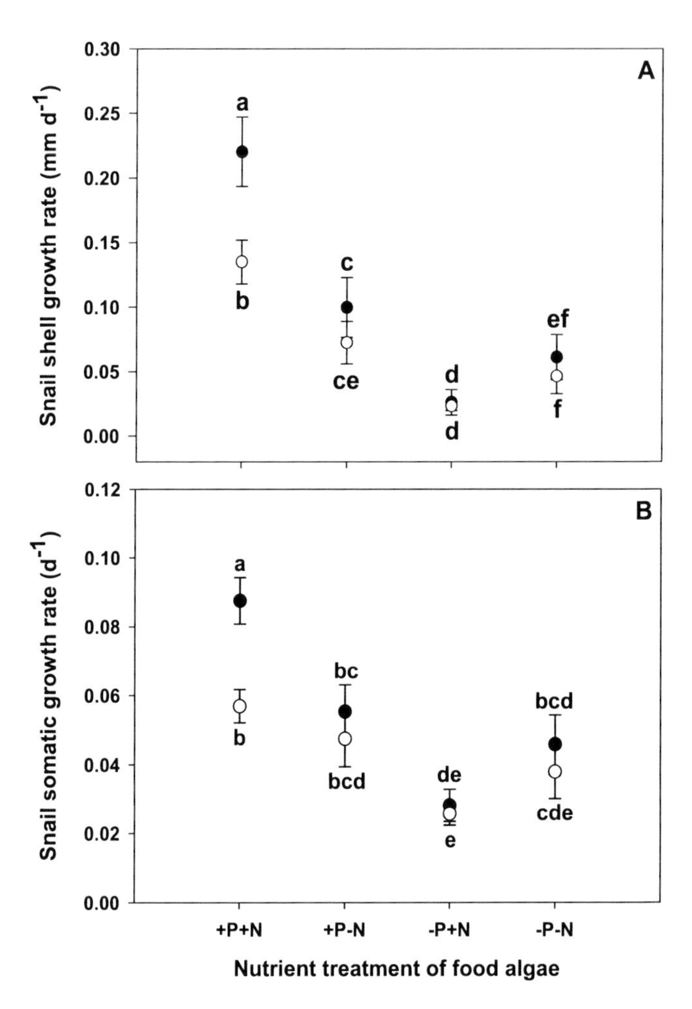

Figure 2: Growth rates of juvenile *R. ovata* in the laboratory experiments. (A) Shell growth rates (mm d^{-1}). (B) Somatic growth rates of snail soft bodies (d^{-1}). Filled symbols, high quantity treatment; open symbols, low quantity treatment. Values are mean ± SD; n=8. Different letters indicate significant differences between treatments (Tukey's HSD multiple comparison test following two-way ANOVA).

In general, the C:P and C:N ratios of the soft bodies of the experimental snails and of the field-collected snails were in a similar range (field: C:P 82 - 238; C:N 4 - 6; experiment: C:P 127 - 390; C:N 4 - 13), and all *R. ovata* C:P and C:N ratios were far below those of the periphyton in the field and of the nutrient-depleted food algae.
In order to test the Growth Rate Hypothesis with *R. ovata*, a linear regression of the soft body growth rates of the experimental animals versus their soft body C:P ratio was determined. A negative relationship (Fig. 4; $R = 0.58$, $p<0.05$) between growth rate and soft body C:P ratio was found, thus supporting the Growth Rate Hypothesis.

Table 2: Results of a two-way ANOVA on *R. ovata* shell and somatic (soft body) growth rates with food quantity and food quality as factors.

		Shell growth rate (mm d^{-1})			Somatic growth rate (d^{-1})		
Factor	d.f.	MS	F	p	MS	F	p
Food quantity	1	0.034	38.29	< 0.001	0.373	124.78	< 0.001
Food quality	3	0.190	211.31	< 0.001	0.554	185.57	< 0.001
Quan x Qual	3	0.006	6.90	< 0.001	0.211	70.61	< 0.001
Error	55	0.001			0.003		

Snail feces C:N:P ratios

Fecal pellets of the experimental animals were collected with a pasteur pipet from the bottom of the containers, and the elemental composition was analyzed on day 25 and day 28 of the experiment. The C:N:P ratios of *R. ovata* feces varied more than that of both snails and their food. The C:P ratios of fecal pellets produced by *R. ovata* fed algae grown with P in the medium were lower than C:P ratios of fecal pellets from snails fed P-depleted algae (Tab. 3). However, the C:N ratios of the feces from snails fed +P-N algae and +P+N algae did not differ (Tab. 3), which reflected the somewhat less clear pattern in the dietary C:N ratios (Tab. 1).
Snails fed P-saturated algae excreted fecal pellets with C:N:P ratios similar to those found in both algae and the soft bodies (Tab. 1, 3, Fig. 3). In contrast to this, C:P ratios of the feces in the -P-N treatment reached very high values of up to almost 3000 when animals were fed P-depleted *U. fimbriata* (Tab. 3). This was significantly higher than in the food algae (two-way ANOVA, $F_{(1, 28)}=5.9$, $p<0.05$), which suggests an efficient assimilation of phosphorus from the diet under P-depleted conditions. However, there was no significant difference between the C:P ratios of food and feces in any of the three other nutrient treatments. Similarly, differences in N:P ratios between food and snail feces were not significant in all treatments except for -P+N

($p<0.05$). Snail feces in the -P+N treatment had higher N:P ratios than the respective food algae, which also indicated an efficient P assimilation from feces.

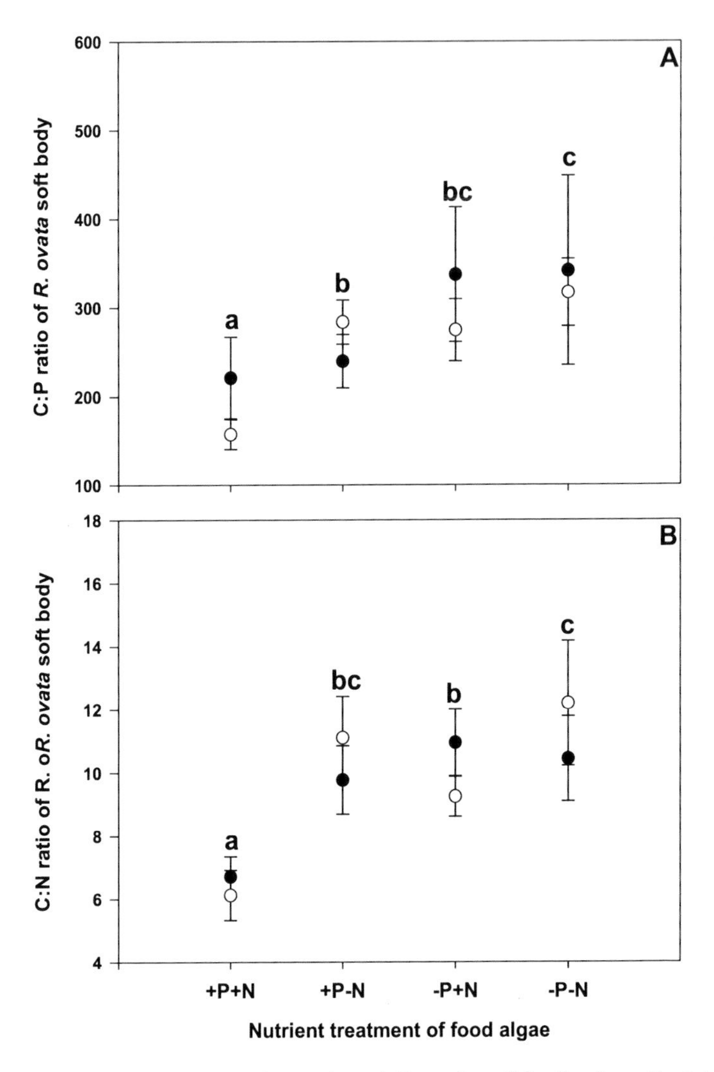

Figure 3: Molar carbon:nutrient ratios of *R. ovata* soft bodies from the laboratory experiments. Values given are means ± SD of n=14. (A) C:P ratios. (B) C:N ratios. Filled symbols, high quantity treatment; open symbols, low quantity treatment.

Food consumption

R. ovata consumed the green alga *U. fimbriata* in all treatments. On day 33, *R. ovata* consumed up to 100% of the food offered in low quantity and max. 93% (57% on average) of the food offered in high quantity, which indicated that food quantity was

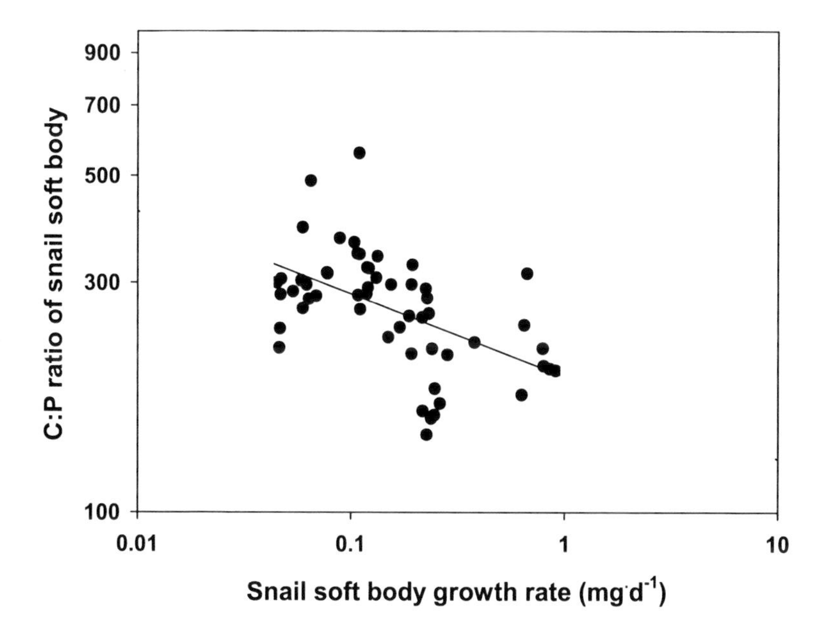

Figure 4: Relationship between *R. ovata* soft body growth rates and the soft body (molar) C:P ratios of the respective individuals. Each dot represents one of 56 analyzed snails. Axes are log/log scaled.

limiting in the +P+N low quantity treatment, but not when food was offered in high quantities in any of the treatments. For comparisons between treatments, the biomass-specific food consumption was calculated to account for the differences in soft body dry weight. A two-way ANOVA revealed highly significant differences in biomass-specific food consumption (mg C mg dry wt.$^{-1}$ d^{-1}), effects of both food quantity ($F=_{(1, 16)}356.4$, $p<0.001$) and quality ($F=_{(3, 16)}23.4$, $p<0.001$), and the interaction of the two factors ($F=_{(3, 16)}37.5$, $p<0.001$). Distinct consumption patterns between treatments were detected using post-hoc comparisons with Tukey's HSD (Fig. 5). In accordance with limitation by food quantity when low amounts of food were offered, consumption per soft body dry weight did not differ between these treatments, regardless of the nutrient regime. There was a highly significant increase in biomass-specific food consumption with the number of mineral nutrients depleted in the algae when high amounts of food were offered. Biomass-specific food consumption of *R. ovata* fed high quantities of *U. fimbriata* in the +P+N treatment did not differ from consumption rates in all but one of the treatments in which low quantities of food were offered. In contrast, food consumption of *R. ovata* fed high quantities of algae grown in the absence of one or two mineral nutrients was higher than when fed nutrient-saturated *U. fimbriata*. Biomass-specific food consumption was highest when high amounts of *U. fimbriata* low in both N and P were fed to *R. ovata*, which resulted in a threefold increase in food consumption compared to that of snails fed the nutrient-saturated algae. This indicates that *R. ovata* is compensating for low-quality food by increasing the food uptake.

Table 3: Mean molar C:N:P ratios of fecal pellets produced by *R. ovata* in the four nutrient treatments. Values given are means (± SD) of two samplings during the experimental period (on day 25 and 28). Letters indicate homogenous groups from Tukey's HSD test following significant effects (C:P $F=24.2$, $p<0.001$; C:N $F=7.8$, $p<0.001$) of a one-way ANOVA on ln(x)-transformed feces nutrient ratios.

Nutrient treatment of the alga	Feces nutrient ratio		Homogenous groups	
	C:P (± SD)	C:N (± SD)	C:P	C:N
+P +N	291 (73)	8 (1)	A	A
+P –N	304 (108)	17 (10)	A	A
–P +N	1501 (1403)	21 (10)	B	AB
–P –N	2965 (1540)	36 (11)	B	B

Discussion

Radix ovata C:N:P ratios and homeostasis

On the one hand, stoichiometric homeostasis, i.e., the maintenance of relatively constant body C:N:P ratios independent of the food source's nutrient status is commonly reported from planktonic (Hessen 1990) and benthic (Frost et al. 2003, Fink et al. *in press*) invertebrates. To maintain this homeostasis, herbivorous zooplankton make physiological adjustments when the food quality (P content) changes (Darchambeau et al. 2003). On the other hand, the view of strict homeostasis of herbivores has been modified recently. The herbivorous zooplankton *Daphnia* sp. can change its somatic C:P ratio (Plath and Boersma 2001) and specific

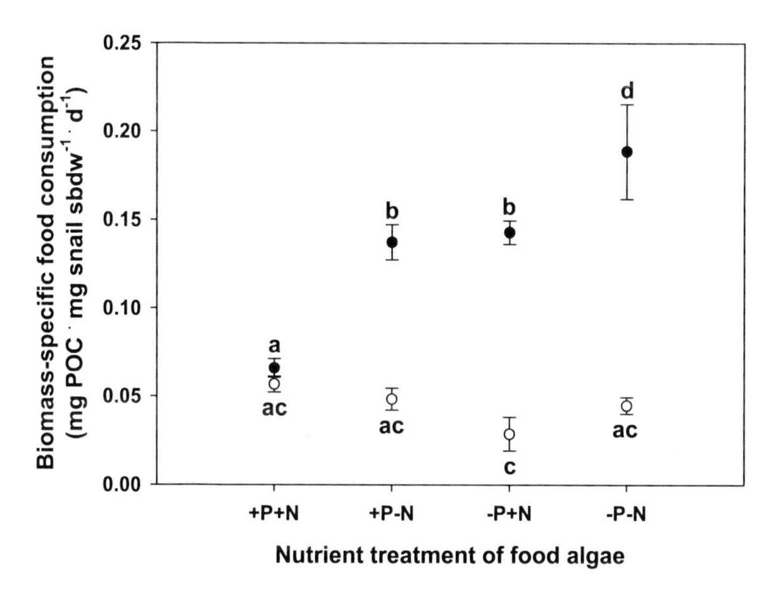

Figure 5: Biomass-specific food consumption (particulate organic carbon dry wt.$^{-1}$ d^{-1}) of *R. ovata* feeding on *U. fimbriata* from different nutrient treatments. Values are mean ± SD; n=3 samples determined on day 33 of the growth experiment. Filled symbols, high quantity treatment; open symbols, low quantity treatment. Different letters indicate significant differences between treatments (Tukey's HSD multiple comparison test following two-way ANOVA).

P content by 0.8–1.6% (of body mass, DeMott 2003), which indicates a limited ability of this freshwater zooplankton to respond to fluctuations in dietary C:P ratios. Whether the benthic herbivore *R. ovata* is a strictly homeostatic consumer and physiologically adapts to maintain this homeostasis, or rather varies its body stoichiometry with changes in the composition of its food, was studied here. Although the C:N:P ratios of the periphyton sampled from the field were highly variable (Fig. 1), the C:N:P ratios of *R. ovata* sampled at the same dates and sites remained constant (Fig. 1). This uncoupling of the C:N:P ratios of the snails and their resource suggests a strongly maintained homeostasis in the herbivore. However, in contrast to the snails sampled in the field, the snails in the laboratory growth experiments differed significantly in C:N:P ratios of their soft bodies (Fig. 3). These differences seemed to be due to a general increase in tissue C:N and C:P ratios when snails were fed nutrient-depleted algae rather than a direct response of the nutrient composition of the soft bodies to the nutrient ratio of the diet. Overall, the P content of *R. ovata* in the growth experiment varied from 0.33 to 0.73% (of body mass). These results indicate that *R. ovata* homeostasis is less strict than suggested from field data. One possible explanation for this could be active selection of food particles of high quality. However, this seems unlikely, as pulmonate gastropods in general, and the genus *Radix* in particular, are considered to be unselectively ingesting the bulk periphyton community (Barnese et al. 1990, Brendelberger 1997a). The observed lower C:P ratios of *R. ovata* grown on nutrient-saturated algae could also be the result of a nutrient storage adaptation to survive periods of low P availability, as has been suggested for mayflies (Frost and Elser 2002a).

DeMott et al. (1998) have found a positive relationship between the specific P content of daphnids and their growth rate when feeding along a gradient of dietary P. Interestingly, such a relationship was also found for *R. ovata*; the snails in the growth experiments showed a negative correlation between the body C:P ratio and growth rate (Fig. 4). This finding is in accordance with the Growth Rate Hypothesis (Elser et al. 2000a), which predicts a higher P demand (and thus, lower body C:P ratio) for fast-growing organisms since they have to maintain high levels of P-rich RNA (Elser et al. 2003).

Snail growth

Stoichiometric constraints for herbivore growth have been extensively studied in zooplankton (reviewed by Sterner and Schulz 1998, Sterner and Elser 2002) and theoretical models on the interacting effects of food quantity and food quality have been derived from those studies (Sterner 1997, Frost and Elser 2002a).

However, both in lentic (Frost and Elser 2002a) and lotic (Stelzer and Lamberti 2002) habitats, only one study to date has demonstrated effects of mineral nutrient limitation on macroinvertebrate growth. The results of the study on mayflies (Frost and Elser 2002a) support model predictions suggesting stronger food quality effects with high quantities of food, whereas investigations with snails (which have higher periphyton removal rates than mayflies (Feminella and Hawkins 1995) and therefore, are probably ecologically more relevant) have found stoichiometry-dependent food quality effects only with low quantities of food (Stelzer and Lamberti 2002). This highlights that disentangling effects of food quantity and food quality on herbivore growth is important, but often difficult (Sterner and Schulz 1998). In our study, we attempted to resolve these conflicting pieces of evidence by comparing limiting with non-limiting food quantities in a factorial design with different resource nutrient treatments, i.e., different food qualities.

Growth rates, measured either as the increase in shell length or in soft body dry weight, indicated that elemental nutrients play a decisive role in herbivore growth. The growth of juvenile *R. ovata* was higher when fed nutrient-saturated algae than when fed nutrient-depleted algae, which demonstrated that the snails in the latter case were P-, N-, or P- and N-limited. This effect of mineral nutrient limitation seemed to be even more important than the effect of food quantity — the shell lengths of snails fed low quantities of nutrient-saturated algae were higher than the shell lengths of snails fed high quantities of nutrient-depleted algae. The results also indicated that feeding on P-depleted algae has more severe consequences than feeding on N-depleted algae for juvenile *R. ovata*. Whether this holds only for juveniles or is valid for all size and age classes of *R. ovata* remains to be tested.

Prior to the experiment, we hypothesized that algae low in both N and P would support even lower growth of *R. ovata* than algae grown in the absence of only one of these nutrients. However, growth of *R. ovata* on *U. fimbriata* grown without both, P and N, was higher than growth of snails fed algae depleted only in P. The reason for this probably is that the growth of *U. fimbriata* had become N-limited before P limitation occurred when both nutrients were absent from the culture medium. Thus, in the absence of both N and P in the algal culture medium, the decrease in the P content of *U. fimbriata* was less pronounced than in the absence of P only. This would lead to higher growth of *R. ovata* on (-P-N) *U. fimbriata*, which corroborates the importance of dietary P.

Strong food quality differences caused by N- and/or P-limitation were observed in the laboratory growth experiments (Fig. 2) at low and especially at high food quantities. The interacting effects of food quantity and food quality and the much higher food quality differences between the high quantity treatments compared to the low quantity treatments support the hypothesis that maximal food quality effects will occur at saturating food quantities (Sterner 1997). In an experiment with a similar design, Stelzer and Lamberti (2002) did not find a food quality effect on the growth of the stream gastropod *Elimia livescens* at high food quantities. One reason for the different findings could lie in the different food and snail species used. Stelzer and Lamberti (2002) fed semi-natural periphyton communities of unknown species composition grown on artificial substrates to the prosobranch snail *E. livescens* in outdoor flow-through mesocosms. We used a pure culture of a benthic green alga (*U. fimbriata*) and a pulmonate gastropod (*R. ovata*) under much more defined laboratory conditions to exclude effects of changes in periphyton community composition caused by differences in the applied nutrient regime. However, our laboratory approach with a monoalgal food is admittedly more artificial than feeding a natural community to gastropod herbivores. The hypothesized lower susceptibility of benthic grazers to energy limitation compared to zooplankton herbivores (Stelzer and Lamberti 2002) is not supported by the highly significant effect of food quantity on *R. ovata* in the present study, which suggests that benthic grazers are susceptible to carbon limitation, and, as a consequence, to interactive effects of food quality and food quantity.

The considerably higher magnitude of the food quality effects on snail growth found in our study compared to the results of Stelzer and Lamberti (2002) was probably due to, at least in part, the magnitude of nutrient depletion in the algal food source. Stelzer and Lamberti fed "low P" periphyton with C:P ratios of 416 to *E. livescens*, whereas we fed highly P depleted algae with C:P ratios of up to 1243 to *R. ovata*; the range in the algal C:P ratios we used is comparable to values found in field-sampled periphyton from Lake Constance (Fig. 1). Our findings suggest that the growth of *R. ovata* is food-quality-dependent and that naturally occurring C:P ratios in periphyton can lead to substantially lower biomass accrual.

Mechanisms to cope with low availability of N or P

What adaptations could benthic herbivores develop to minimize the effects of low N or P availability in their diets? Zooplankton increase the assimilation efficiency for the nutrient limiting in the diet (DeMott et al. 1998). The results of our study indicated that this holds true also for the freshwater snail *R. ovata*. It remains to be tested whether differential excretion of non-limiting nutrients has a feedback on the periphyton community, as suggested by Hillebrand et al. (2004).

Apart from an increased assimilation or an active selection of food particles of high quality (which is unlikely for Radix, see Brendelberger 1997a), an alternative mechanism to cope with low food quality is compensatory feeding, i.e., an increase in food consumption (per grazer biomass) as a response to food deficient in essential compounds. Compensatory feeding responses have been found in different groups of organisms (terrestrial insects, Raubenheimer 1992, and marine amphipods, Cruz-Rivera and Hay 2000) and various ecosystems , but mostly in connection with essential amino acids. To date, only few studies have investigated the potential of compensatory feeding to reduce stoichiometric constraints of herbivore growth and the results were somewhat contradictory. DeMott et al. (1998) and Darchambeau et al. (2003) working with zooplankton (*Daphnia* sp.), and Stelzer and Lamberti (2002) working with a benthic herbivore (*Elimia* sp.) were not able to show such a compensatory feeding response. However, Plath and Boersma (2001) suggested that an increase in appendage beat rates of *Daphnia magna* in response to a P-depleted diet was evidence for compensatory feeding. Here, we found an up to threefold increase in biomass-specific food consumption when *R. ovata* was fed high quantities of nutrient-depleted *U. fimbriata*. When fed high quantities of algae, the biomass-specific consumption was higher when the algae were depleted in P and/or N. Snails that were fed low quantities of algae consumed all the offered food and, therefore, a compensatory feeding response was not possible.

One problem inherent to a compensatory feeding response to low N or P is that the feeder must get rid of the excess carbon acquired during the increased food uptake (Sterner and Hessen 1994). Several mechanisms to cope with excess C have been suggested. Herbivores could lower their uptake efficiency for C (DeMott et al. 1998), increase their C storage (e.g., as lipid droplets, Tessier and Goulden 1982), or increase either C excretion or respiration rates (Darchambeau et al. 2003). To date, such studies have been performed exclusively with daphnids. Here, the increased C:P ratios of the feces excreted by *R. ovata* fed N- and P-depleted algae (Tab. 3) suggested that either an increase in P uptake efficiency and/or an increased excretion of C (relative to P) in the fecal pellets constitutes such an adaptation to surplus dietary carbon in this freshwater snail. If other adaptive mechanisms such as carbon storage in lipids or increased respiration rates also apply for *R. ovata*, remains to be investigated.

Ecosystem consequences

Freshwater gastropods in general and the genus *Radix* in particular are important links for the functioning of freshwater littoral food webs. They are the herbivores with the highest individual biomass, and therefore, individual grazing impact, and they can reach very high abundances, resulting in dominance among benthivorous invertebrates, e.g., in the pre-alpine Lake Constance (Baumgärtner 2004). Therefore, constraints in the biomass accrual of these organisms can be expected to have considerable impact on the energy transfer at the plant–herbivore interface in benthic communities and therefore also on higher trophic levels. However, the role of food-quality constraints on the energy transfer efficiency between trophic levels in freshwater benthic systems has been largely neglected. The results of our study highlight that ecological stoichiometry is not only of high relevance for terrestrial and planktonic freshwater ecosystems (Elser et al. 2000b), but also for freshwater benthos.

Even though *R. ovata* does not seem to be strictly homeostatic, low availability of N and especially P strongly constrained the growth of the snails, resulting in considerably lowered herbivore biomass accrual on nutrient-limited periphyton. This suggests that even for non-strictly homeostatic consumers, a low dietary supply of P and N can have severe consequences, and it is expected that stoichiometric limitation of primary production frequently found in a variety of ecosystems can propagate through benthic food webs and affect all trophic levels. The pronounced compensatory feeding response of *R. ovata* could partly dampen these food quality effects. However, the effects of nutrient availability on snail biomass accrual were still considerable. Increased feeding of gastropods at low availability of P and N can lead to depletion of periphyton biomass, and hence compensatory feeding might ultimately shift the benthic herbivore community from a P or N limitation to a C limitation.

Theoretically, in ecosystems with high grazing and/or low productivity, snails could become limited by food quantity (i.e. carbon availability) caused by their own compensatory feeding response. However, this is not necessarily the case in Lake Constance, where periphyton biomass is never depleted beyond 0.4 mg ash-free dry mass cm^{-2} (Fink et al. *in press*). Probably, productivity of Lake Constance periphyton was too high to result in a food quantity limitation, despite high grazer abundance during the summer months (Baumgärtner 2004). High grazing pressure and low periphyton production commonly co-occur both in marine and freshwater littoral zone habitats (Hillebrand and Kahlert 2001). This exerts a strong selection pressure on herbivores to evolve strategies to cope with low nutrient availability, e.g. compensatory feeding – this, in turn will influence ecosystem structure and productivity.

Acknowledgements

We are indebted to R. S. Stelzer for valuable comments that greatly improved an earlier draft of this manuscript and to K. A. Brune for editing the English. We thank C. Gebauer and P. Merkel for technical assistance with the nutrient analyses and S. Boekhoff for help with the experiments. This study was supported by the Deutsche Forschungsgemeinschaft (DFG) within the Collaborative Research Centre SFB 454 – Littoral of Lake Constance.

Chapter 4

Food quality of algae and cyanobacteria for the freshwater gastropod *Bithynia tentaculata*: the role of polyunsaturated fatty acids

Patrick Fink & Eric von Elert

Introduction

In aquatic food webs, the energy transfer between primary producers and herbivores is often found to be quite variable. For the understanding of these food webs, it is crucial to identify key factors that determine the transfer efficiency at the plant-herbivore interface. One parameter strongly affecting the transfer efficiency is the nutritional quality of the ingested food items. Food quality of algae and cyanobacteria for herbivores can be determined by a multitude of different factors such as cell morphology (Porter 1975, Van Donk et al. 1997), the content of mineral nutrients (reviewed by Sterner and Elser 2002) or by biochemical constituents. Among these biochemical constituents, toxins (e.g., Lampert 1981) as well as the lack of essential biochemical compounds such as polyunsaturated fatty acids (PUFAs, Müller-Navarra 1995, Wacker and von Elert 2001) or sterols (Von Elert et al. 2003) can lead to reduced fitness of herbivores. Especially the group of n-3 PUFAs are considered to be essential for many invertebrates (Stanley-Samuelson et al. 1988). Among these n-3 PUFAs, α-linolenic acid (α-LA, C18:3 n-3) and eicosapentaenoic acid (EPA, C20:5 n-3) have been shown to play an important role for the growth of the herbivorous zooplankton species *Daphnia* (Müller-Navarra 1995, Wacker and von Elert 2001). However, until now, relatively little is known about the role of PUFAs in determining food quality effects in freshwater benthic organisms. Only recently, thorough investigations on molluscs have been performed. These have demonstrated a potentially limiting role of PUFAs in determining food quality, as the PUFA content of algae influenced different life history stages of the bivalve *Dreissena polymorpha* (Wacker 2002). Another group of molluscs, the gastropods are important benthic herbivores in many freshwater littoral habitats (Feminella and Hawkins 1995). They reach considerably high abundance and biomass and play a major role in structuring the algal communities of littorals (Cattaneo and Kalff 1986, Feminella and Hawkins 1995). Therefore, it can be assumed that food quality effects on gastropod grazers can have high impact on the structure of freshwater littoral food webs. The prosobranch gastropod *Bithynia tentaculata* is among the most common species in central European lakes and is able to feed selectively on high quality food (Brendelberger 1997a). To investigate the food quality of primary producers for freshwater gastropods and the potential role of polyunsaturated fatty acids (PUFAs) in determining food quality differences, a growth experiment with juvenile *Bithynia tentaculata* was conducted under controlled laboratory conditions in which the n-3 PUFAs α-LA and EPA were added as single compounds or in combination to a cyanobacterium and two species of freshwater planktonic algae. Juvenile growth rates determined as increase in shell length as well as gain in soft body dry mass were considered as indicators of food quality.

Methods

Food organisms

Suspensions from unialgal cultures of a green alga (*Scenedesmus obliquus*, SAG 276-3a), a diatom (*Cyclotella meneghiniana*, SAG 1020 – 1a) and a cyanobacterium (*Chroococcus minutus*, SAG 41.79), all originating from the Sammlung für Algenkulturen, Göttingen, Germany, were offered as food sources to *B. tentaculata*. All food organisms were grown at a temperature of 20° C in semicontinous batch cultures in different growth media and a light intensity of 120 $\mu E \cdot s^{-1} \cdot cm^{-2}$. *S. obliquus* was grown on a modified WC medium [(Von Elert 2002), modified from (Guillard 1975)], *C. minutus* on Cyano medium (Jüttner et al. 1983) and *C. meneghiniana* on Diatom medium (Wendel 1994). The green alga *S. obliquus*, known to contain high amounts of α-LA (C18:3 n-3) but no EPA (C20:5 n-3, Von Elert 2002) was supplemented with EPA using the method of (Von Elert 2002). In the same way, *C. meneghiniana*, which contains relatively high amounts of EPA, but almost no α-LA (Wacker and von Elert 2002) was supplemented with α-LA. The cyanobacterium *C. minutus* contains no PUFAs. Therefore, apart from the unsupplemented cyanobacteria, snails were also fed *C. minutus* supplemented either with α-LA, EPA or with both n-3 PUFAs. The *B. tentaculata* growth experiment was performed with eight treatments, three unsupplemented suspensions of the primary producers and five with addition of either α-LA, EPA or both PUFAs. All food suspensions were adjusted to a carbon concentration of 0.25 mg C · ml^{-1} by photometric light extinction (800 nm) using carbon extinction equations.

Fatty acid analyses

Fatty acids in supplemented and unsupplemented algae and cyanobacteria were determined from subsamples of the food suspensions using the method described by Von Elert and Stampfl (2000). PUFAs were quantified as fatty acid methyl esters (FAMEs) using a HP 6890 GC (Agilent Technologies, Waldbronn, Germany) equipped with a DB 225 fused silica column (J&W Scientific, Folsom, USA) and a flame ionisation detector with heptadecanoic acid methyl ester and tricosanoic acid methyl ester as internal standards as Von Elert and Stampfl (2000). Identification of the FAMEs was based on comparison of retention times to those of reference compounds. For details on the method see Von Elert (2002).

Growth experiment

B. tentaculata has a 2-3 year life cycle with the main growth phase during the first summer (Brendelberger 1995a), therefore the juvenile growth rate was measured as a fitness parameter. For the calculation of this growth rate, soft body dry mass (SBDM) is considered to be the better indicator of snail growth than shell length, because e.g. during periods of starvation the soft body is reduced while the shell length remains constant (Brendelberger 1995a). Therefore, juvenile *Bithynia tentaculata* (shell length ~ 3.5 mm) collected from the littoral of Lake Constance were fed over a 21-day period with an amount of the respective food alga equivalent to 0.5 mg C · d^{-1} per individual. Each treatment was replicated eightfold, resulting in a total number of 64 experimental units. Snails were kept individually in polypropylene boxes with filtered (0.45 µm) Lake Constance water at 20° ± 0.5° C and constant dim light (20 µE s cm^{-2}). Food and water were renewed every other day and containers replaced twice per week. *B. tentaculata* is not only able to graze on benthic biofilms, but can also switch to a second mode of feeding, filtration of phytoplankton with their gills (Brendelberger and Jürgens 1993). Therefore, the food organisms were suspended in the experimental containers in order to allow the animals both gill filtration before and grazing after sedimentation of the food particles. Snail growth was recorded both as increase in shell length and gain in SBDM. Shell length of every individual was determined weekly using a dissecting microscope equipped with a digital image analysis system. SBDM was determined at the end of the experiment by dissecting the soft bodies under a microscope and measuring the mass of freeze-dried soft bodies. Start values of SBDM were calculated from the initial shell lengths using a shell length/SBDM regression of field-collected *B. tentaculata* (data not shown). Growth per day for shell length and SBDM was calculated and it was tested for differences between treatments using one-way ANOVA with post-hoc Tukey HSD comparisons (Statistica'99, StatSoft Inc.).

Results

Fatty acid supplementation

Unsupplemented *Chroococcus minutus* contained no PUFAs at all (Tab. 1). Only saturated and monounsaturated fatty acids with up to 18 carbon chain length were found to occur naturally in this cyanobacterium which is in accordance with the assumed incapability of cyanobacteria to synthesize long-chained and polyunsaturated fatty acids (Cobelas and Lechardo 1988). In contrast to this, the diatom *Cyclotella meneghiniana* exhibited a diverse array of long chained polyunsaturated fatty acids of both the n-6 and the n-3 type (Tab. 1, see also Wacker and von Elert 2002). Both α-LA and EPA were found to occur naturally in *C. meneghiniana*, however the natural α-LA content being very low compared to the EPA content. Unsupplemented cultures of the green alga *Scenedesmus obliquus* contained considerable amounts of the C18 PUFAs linoleic acid, α-linolenic acid and traces of stearidonic acid, but no fatty acids > C_{18} like eicosapentaenoic acid (Tab. 1, see also Von Elert 2002). Supplementation of *C. minutus*, *C. meneghiniana* and *S. obliquus* with the n-3 PUFAs α-LA and EPA acid resulted in pronounced increases of these PUFAs in the cells of all these primary producers (Tab. 1).

Table 1: Fatty acid composition of the cyanobacterium and the algae used as food sources for *Bithynia tentaculata* before and after supplementation with either α-linolenic acid (C18:3 n-3), eicosapentaenoic acid (C20:5 n-3) or both n-3 PUFAs. The values are means of n=3 (*C. minutus* + α-LA and *S. obliquus* + EPA n=1) fatty acid extractions and measurements (n.d. = not detected), the standard deviation is given in parentheses.

	Fatty acid content (µg mg C^{-1})							
fatty acid	*C. minutus*	*C. minutus* + α-LA	*C. minutus* + EPA	*C. minutus* + α-LA + EPA	*C. mene-ghiniana*	*C. mene-ghiniana* + α-LA	*S. obliquus*	*S. obliquus* + EPA
C14:0	6.2 (3.6)	4.9	3.8 (0.3)	4.3 (1.8)	12.9 (3.6)	12.4 (1.0)	1.4 (1.2)	1.8
C14:1	4.4 (3.4)	3.1	2.1 (0.2)	2.6 (1.4)	n.d.	n.d.	n.d.	n.d.
C15:0	n.d.	n.d.	n.d.	n.d.	3.9 (0.4)	3.8 (0.3)	n.d.	n.d.
C16:0	7.0 (0.3)	6.6	6.5 (0.3)	6.7 (0.5)	81.2 (4.2)	78.8 (4.1)	10.9 (9.6)	14.3
C16:1	1.2 (0.0)	n.d.	n.d.	n.d.	122.5 (8.5)	113.4 (9. 8)	0.7 (0.6)	0.9
C18:1 n-9/n-12	3.4 (0.0)	3.3	3.3 (0.0)	3.3 (0.3)	2.3 (0.9)	0.6 (1.1)	12.4 (10.7)	18.1
C18:1 n-7	n.d.	n.d.	n.d.	n.d.	1.2 (0.1)	0.3 (0.5)	0.7 (0.6)	n.d.
C18:2 n-6	n.d.	n.d.	n.d.	n.d.	1.4 (0.3)	0.5 (0.8)	8.7 (7.5)	9.1
C18:3 n-6	n.d.	n.d.	n.d.	n.d.	1.2 (0.09)	0.3 (0. 6)	0.9 (0.8)	1.0
C18:3 n-3	n.d.	11.4	n.d.	9.4 (1.6)	1.3 (0.06)	47.1 (6.4)	25.9 (22.4)	27.3
C18:4 n-3	n.d.	n.d.	n.d.	n.d.	4.6 (0.1)	5.4 (0.7)	2.9 (2.6)	3.1
C20:5 n-3	n.d.	n.d.	6.1 (2.3)	5.7 (1.5)	28.8 (1.3)	27.0 (4.0)	n.d.	25.7

However, the efficiency of the supplementation method was lower for the cyanobacterium *C. minutus* both with α-LA and EPA than with the eukaryotic algae. The content of other fatty acids remained largely unchanged by the supplementation with n-3 PUFAs.

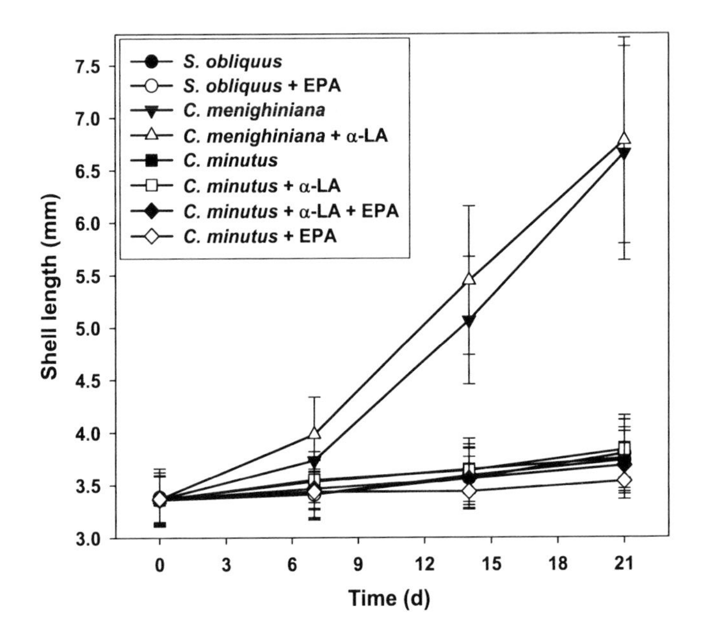

Figure 1: Mean shell length of juvenile *Bithynia tentaculata* fed algae and cyanobacteria with and without previous supplementation of PUFAs over a 21 d experimental period at 0.5 mg POC per ind. and day over a 21 d period at 20° C. Values are mean ± SD, n=8.

Bithynia growth

When juvenile *B. tentaculata* were grown for 21 days on *C. meneghiniana*, *S. obliquus* and *C. minutus* with and without supplementation of n-3 PUFAs, one-way ANOVA on snail growth parameters showed highly significant differences both in terms of total shell length ($F = 51{,}03$, $p < 0{,}001$, Fig. 1) and somatic growth rates calculated as gain in SBDM per individual and day ($F = 35{,}18$, $p < 0{,}001$, Fig. 2). *C. meneghiniana* led to significantly higher growth, indicating higher food quality for *B. tentaculata* than *S. obliquus* and *C. minutus*.

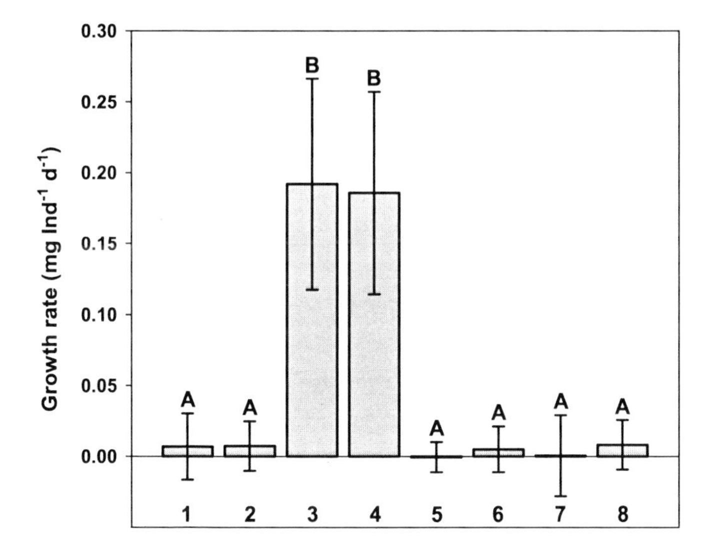

Figure 2: Juvenile growth rates of *Bithynia tentaculata* fed either the green alga *Scenedesmus obliquus* (1), *S. obliquus* + C20:5n-3 (2), the diatom *Cyclotella meneghiniana* (3), *C. meneghiniana* + C18:3n-3 (4), the cyanobacterium *Chroococcus minutus* (5), *C. minutus* + C18:3n-3 (6), *C. minutus* + C18:3n-3 + C20:5n-3 (7) or *C. minutus* + C20:5n-3 (8) calculated as mg increase in soft body dry mass per individual and day. Algae were fed at 0.5 mg POC per individual and day over a 21 d period at 20° C. Values are mean ± SD, n=8. Different capital letters indicate significant differences between treatments (Tukey's HSD multiple comparison test following one-way ANOVA).

However, this difference was not affected by supplementation with either of the PUFAs. Addition of α-LA to *C. meneghiniana* led to a slight but not significant increase in *B. tentaculata* shell length during the second week of the experiment (Fig. 1, F = 1,24, p = 0,28). Supplementation of *S. obliquus* with EPA did not change its food quality and neither did addition of either α-LA or EPA as single compounds or in combination to *C. minutus* (Fig. 1, 2).

Discussion

Large differences in food quality between the algal and cyanobacterial species fed to *B. tentaculata* were found. Such differences in food quality can have a multitude of

reasons. First, the morphology of the cells could play an important role. Colonial or filamentous forms can be as difficult to ingest due to their large size (Cattaneo and Kalff 1986) as very small cells that could for example go through filter meshes during the gill filtration or through the interspaces between the radular teeth (Brendelberger 1997a). Thick cell walls or mucilaginous sheaths can make cells resistant to digestion (Porter 1975, Van Donk et al. 1997). But even if food items are ingestible and digestible, there is often a dietary mismatch between the essential contents of the food item and the needs of the consumer, which leads to low food quality even at a high availability of food quantity (Sterner and Schulz 1998). One aspect of food quality discussed frequently in freshwater ecology is the elemental stoichiometry of carbon, nitrogen and phosphorous (Sterner and Elser 2002). However, in the food quality experiment of this study, it was attempted to exclude effects of an inadequate elemental composition by using semicontinous culture in nutrient-rich algal media. Another aspect of a dietary mismatch apart from elemental stoichiometry is a lack of essential biochemical compounds, such as polyunsaturated fatty acids (PUFAs). Especially the group of n-3 PUFAs, that can be synthesized by plants and algae but not by most animals (Stanley-Samuelson et al. 1988) have shown to be limiting herbivore growth in freshwater ecosystems (e.g., Müller-Navarra 1995, Wacker and von Elert 2001, Wacker and von Elert 2002). The common freshwater gastropod species *B. tentaculata* was chosen as a model organism for this study because it can both feed on suspended food particles as well as on algae attached to surfaces (Brendelberger and Jürgens 1993). Therefore, the method of Von Elert (2002) to supplement planktonic algae with single fatty acids could be applied here to a benthic organism without introducing further problems with the supplementation of food organisms attached to hard substrates. Furthermore, *B. tentaculata* was previously found to be able to feed selectively on high quality food (Brendelberger 1997a).
Although the diatom *C. meneghiniana* contains only traces of α-LA, it could be well assimilated by *Bithynia* as is evidenced by the growth of the snails and supported by data from the literature, suggesting higher food quality of diatoms for *B. tentaculata* than of green algae and cyanobacteria (Brendelberger 1997a). Supplementation of *C. meneghiniana* with α-LA did not lead to a further increase of snail growth. During the second week of the experiment, snails feeding on *C. meneghiniana* supplemented with α-LA seemed to grow slightly better, however this effect was weak and not statistically significant.

Growth of *B. tentaculata* on both the green alga *S. obliquus* and the cyanobacterium *C. minutus* was very low compared to growth on *C. meneghiniana*. Neither presence nor absence of α-LA nor EPA affected the nutritional value of these food organisms for the snails. According to Brendelberger and Jürgens (1993) *Bithynia* filtration rates are independent of food size, motility and cell surface. This suggests that the observed low growth on *S. obliquus* and *C. minutus* was rather due to low digestibility and/or assimilation than to interference with the ingestion process. This is corroborated by the occurrence of faecal pellets in the *S. obliquus* treatments, indicating that the relatively thick cell walls of *Scenedesmus* (Van Donk et al. 1997) resisted digestion and thus led to low growth of *B. tentaculata*. As *S. obliquus* was probably not usable as food source independent of the biochemical contents, it remains to be tested, what constrains the quality of green algae with cell walls less rigid than those of the *Scenedesmus* strain used in this study.
As the cyanobacterium *C. minutus* does not possess a rigid cell wall, its low nutritional quality for *Bithynia* could be caused either by its small cell size (but see Brendelberger and Jürgens 1993), by the presence of some type of toxin like it is frequently found in different types of cyanobacteria (Lampert 1981) or by a lack of some essential compound limiting the snails' growth on this food source. Toxicity can probably be excluded, as almost no mortality occurred during the whole experimental period and *Chroococcus* has so far not been reported to produce any toxins. *B. tentaculata* feeding on *C. minutus* produced faecal pellets (though less than on the diatom or the green alga). This suggests that the low growth of *B. tentaculata* on *C. minutus* is not due to interference with the ingestion but rather with the assimilation of the cyanobacterial carbon by the snails. As PUFAs were also not sufficient in explaining the nutritional inadequacy of cyanobacteria for *Daphnia* (Von Elert and Wolffrom 2001), *Bithynia* might be constrained by the lack of some other compound like sterols (Von Elert et al. 2003) when confronted with this cyanobacterium as single food source. This remains to be tested in further experiments.
In conclusion, the result of this experiment with *Bithynia tentaculata* demonstrates, that neither the absence of C20:5n-3 (*S. obliquus*) nor a relatively low content of C18:3n-3 (*C. meneghiniana*) constrained food quality for *B. tentaculata*. Hence, not withstanding results obtained with *Daphnia* (Wacker and von Elert 2001) or the bivalve *Dreissena* (Wacker and von Elert 2002), it seems that PUFAs in general and especially α-LA and EPA are not among the main limiting resources for the growth of

this freshwater gastropod. It remains however, to be tested if the reproduction of this mollusc can be influenced by the PUFA content of the spawning adults' food (Wacker and von Elert 2003, Wacker and von Elert 2004).

Acknowledgement

This work was supported by the German Research Foundation (DFG) within the Collaborative Research Centre SFB 454 "Littoral of Lake Constance".

Chapter 5

You are what you eat - Growth and fatty acid composition of the freshwater gastropod *Radix ovata* fed on algae with natural and modified fatty acid patterns

Patrick Fink and Eric von Elert

Abstract

The nutritional quality of algae for herbivore consumers determines how efficiently photosynthetically fixed energy is transferred through food webs. In freshwater ecosystems, there is accumulating evidence, that a major factor in determining this nutritional quality of algae is the content of polyunsaturated fatty acids (PUFAs). In particular, the group of n-3 PUFAs appears to be essential for most animals. Therefore, the algal content of these n-3 PUFAs was found to be a good predictor of the nutritional quality of these algae for herbivores. However, the importance of PUFAs for algal food quality was only investigated with filter-feeding herbivores and phytoplankton. Here, we investigated the role of dietary PUFAs for the growth of juvenile freshwater gastropods in a laboratory growth experiment. Four benthic algal taxa differing in their PUFA composition were fed in equal amounts to juvenile snails of the species *Radix ovata*. Furthermore, a diatom and a green alga that contain different n-3 PUFAs were artificially supplemented with n-3 PUFAs not naturally occurring in these species, to investigate the specific role of particular n-3 PUFAs for the growth of *R. ovata*. Although *R. ovata* readily ingested all the offered food organisms, differences in the algae's nutritional quality measured as the growth increment of the snails were pronounced. Neither the total fatty acid content nor the content of n-3 PUFAs of the algae influenced the nutritional value of these algal taxa for *R. ovata*. Similarly, the addition of specific single PUFAs had no effect on *R. ovata* growth rates. However, the diet's fatty acid composition strongly influenced the fatty acid composition of the herbivores themselves. Specific PUFAs strongly increased in *R. ovata* soft bodies, when the snails were fed a diet with a high content of these PUFAs. Furthermore, *R. ovata* appears to have considerable abilities for the truncation, elongation, saturation and desaturation of dietary fatty acids. This high biosynthetic ability probably explains why *R. ovata* was not found to be limited by the availability of PUFAs. This might be a mechanism for this benthic herbivore to cope with fluctuating availabilities of dietary PUFAs in the field.

Keywords: Eicosapentaenoic acid, food quality, freshwater gastropod, *Gomphonema parvulum*, growth rate, α-linolenic acid, PUFA, *Radix ovata*, *Ulothrix fimbriata*

Introduction

The plant-herbivore interface is the crucial step in food webs to determine how efficiently primary production is transferred to higher trophic levels. However, the efficiency of herbivores to utilize this source of photosynthetically fixed energy is highly variable. Therefore, the factors determining the efficiency of this energy transfer have been in the focus of attention of research on trophic interactions. Apparently, not all food items are equally good in supporting growth of their consumers. Thus, these differences not determined by the food quantity are defined as food quality differences (Sterner and Schulz 1998). The mineral nutrients nitrogen and phosphorous are known to play a major role in determining the food quality of algae for zooplankton. For example, phosphorous-limited algae are known to support significantly lower herbivore growth than algae grown without nutrient limitation (Sterner 1993, Fink and Von Elert *submitted*). The role of mineral nutrients has been investigated in great detail during the last decade. More recently, research in food quality of algae for freshwater plankton focussed on a food quality constraint other than mineral nutrients. For the planktonic herbivore *Daphnia*, it seems that, when molar C:P ratios in the seston are below a value of 300 (Sterner and Schulz 1998), the availability of polyunsaturated fatty acids (PUFAs) becomes the major constraint for *Daphnia* growth (Müller-Navarra 1995, Wacker and von Elert 2001). Similarly, the availability of PUFAs has a strong effect on different ontogenetic stages in the life cycle of the zebra mussel *Dreissena polymorpha* (Wacker et al. 2002, Wacker and von Elert 2002, Wacker and von Elert 2003, Wacker and von Elert 2004). Especially the group of n-3 PUFAs seems to be important, as this group cannot be synthesised by invertebrates (Stanley-Samuelson et al. 1988). Hence, it seems, that PUFAs can play a major role in determining energy transfer efficiency at the algae-herbivore interface of freshwater food webs (Brett and Müller-Navarra 1997). However, the role of PUFAs for the growth of freshwater herbivores has been intensively studied in zooplankton (Von Elert 2002) or in the filter-feeding bivalve *Dreissena polymorpha* (Wacker and von Elert 2002), but not for benthivorous organisms. The pulmonate gastropod *Radix ovata* (DRAPARNAUD) feeds exclusively on benthic food sources, e.g. epilithic (attached to hard substrates) communities in the littoral zone. These benthic algal communities show strong variability in taxonomic (Swamikannu and Hoagland 1989, Stevenson et al. 1996) and nutrient composition (Kahlert 1998, Fink et al. *submitted*). The growth of *R. ovata* can be constrained by the availability of mineral nutrients (Fink and Von Elert *submitted*). However, like for zooplankton herbivores (Sterner and Schulz 1998), when mineral nutrients are available in sufficient amounts, other cell constituents such as polyunsaturated fatty acids (PUFAs) might become limiting for this gastropod herbivore.

In this experiment, we investigated, whether the availability of the n-3 PUFAs α-linolenic acid (ALA, C18:3 n-3) and eicosapentaenoic acid (EPA, C20:5 n-3) determines the food quality of different algal species for the growth of the freshwater gastropod *Radix ovata*. The method developed by Von Elert (2002) was used to modify the fatty acid patterns of two species of primary producers. We hypothesised that (i) the PUFA composition of benthic algae commonly occurring in lake periphyton communities has an influence on the growth of *R. ovata* and (ii) the PUFA composition of the food algae could not only determine the algae's nutritive value for *R. ovata*, but will also affect the PUFA composition of the snail's soft body.

Methods

Food cultures

As food algae for *R. ovata*, pure cultures of two benthic diatoms and two filamentous green algal species were used. Benthic diatoms and green algae are the most important components of periphyton communities in many lake littoral zones (Stevenson et al. 1996) and are known to be readily ingested by gastropod grazers (Swamikannu and Hoagland 1989). The green algae *Ulothrix fimbriata* (BOLD) SAG 36.86 and *Zygnema circumcarinatum* (CZURDA) SAG 698-1a and the diatom *Gomphonema parvulum* (KÜTZING) SAG 1032-1 were obtained from the Sammlung für Algenkulturen Göttingen, Germany. Furthermore, a strain of the benthic diatom *Achnanthidium* sp. isolated from cobblestones in the Lake Constance littoral zone by O. Neuschäfer-Rube, was used as food organism for *R. ovata*.

The green algae were cultivated on WC medium (Guillard and Lorenzen 1972) and the diatoms were kept on Diatom medium (Wendel 1994). All algal cultures were cultivated semicontinously at a dilution rate of 0.25 d^{-1} at 20° C with a light intensity of $1 \cdot 10^{16}$ quanta s^{-1} cm^{-2}. Every other day, algae were harvested by centrifugation at 4000 x g and adjusted to a carbon concentration of 0.25 mg C $\cdot$ ml^{-1} by using previously determined carbon extinction equations at 800 nm.

Experimental animals

Juvenile *Radix ovata* (DRAPARNAUD), a pulmonate gastropod periphyton grazer (Calow 1970) were sampled at approx. 40 cm depth in the littoral zone of Lake Constance, where they are the most abundant benthivorous invertebrate (with respect to biomass, Baumgärtner 2004). Juveniles were chosen for the growth experiment as they have higher specific growth rates than adults and should therefore be more susceptible to food-quality constraints for growth (Sterner and Schulz 1998). Shell lengths of the field-collected juvenile *R. ovata* were determined to the nearest 0.01 mm within 24 hours of sampling using a dissecting microscope equipped with a digital camera and an image analysis system. Here, the shell length is defined as the distance from the apex to the most distal part of the shell's aperture.

Enrichment of algal suspensions with single fatty acids

To enrich suspensions of algae with single fatty acids, the method established by Von Elert (2002) for planktonic algae was used. However, it was not possible to supplement the benthic algae while they were attached to substratum. Therefore, by applying heavy aeration of the culture vessels, we forced the benthic algae into a „semi-planktonic“ culture of which only the suspended part was used for the supplementation. Thus, for the suspended algal cells, the method for planktonic algae could be applied. The green alga *U. fimbriata*, known to contain high amounts of ALA (C18:3 n-3), but no PUFAs with chain lengths of more than C_{18} and thus, no EPA (C20:5 n-3), was supplemented with EPA. In the same way, *G. parvulum*, which contains relatively high amounts of EPA, but no ALA, was supplemented with ALA.

Experimental setup

The growth experiment was performed with six treatments, four unsupplemented algal suspensions and two with addition of either ALA or EPA. Each treatment was replicated fivefold, resulting in a total of 30 experimental units. During the experiments, the snails were kept individually in 400-ml polyethylene containers (Buchsteiner, Gingen, Germany) containing 200 ml of 0.45 µm filtered Lake Constance water. The containers were kept at 20 ± 0.5°C under dim light to minimize algal growth. Every other day, water, faeces, and remaining food were replaced with freshly filtered Lake Constance water and new food suspensions were added at a biomass equivalent to 1 mg C $Ind.^{-1}$ d^{-1}. The snails were transferred to new containers once every week. Snail growth was recorded both as increase in shell length and gain in soft body dry weight (SBDW). Shell length of every individual was determined every 10 d as described above. At the end of the experiment (after 41 days) the soft bodies were removed from the shells under a dissecting microscope and frozen immediately at –80° C. Frozen soft bodies were freeze-dried and the SBDW was determined with a microbalance (Mettler UTM2) to the nearest microgram. Snail growth rates were determined as the incremental increase in SBDW (mg dry wt. d^{-1}) from the beginning of the experiment until the end. For the calculation of snail growth rates, SBDW is considered to be the better indicator of snail growth than shell length, because, e.g. during periods of starvation, the soft body is reduced while the shell length remains constant (Brendelberger 1995a). After dry weight determination, the soft bodies were extracted for fatty acid analyses.

Shell lengths and soft body dry weights of 24 individuals remaining from the initial field sampling were removed from their shells, freeze dried, and stored at –80°C. These animals were used to establish a linear regression of shell length versus soft body dry weight (R^2 = 0.89). The initial soft body dry weight of the snails in the growth experiment could not be determined directly, therefore, the initial soft body dry weights were calculated using the regression and the shell lengths measured at the start of the experiment.

Elemental analyses

To analyse C:N:P ratios of the algal cultures, an aliquot of each algal suspension was filtered on a precombusted glass-fibre filter (Whatman GF/F , Whatman, Maidstone, UK) and dried for subsequent analysis of particulate organic carbon and particulate organic nitrogen with an NCS-2500 analyzer (Carlo Erba Instruments). For determination of particulate phosphorous, an aliquot of each algal suspension was filtered through acid-rinsed polysulfone membrane filters (HT-200, Pall, Ann Arbor, Mich., USA) and digested with a solution of 10% potassium peroxodisulfate and 1.5% sodium hydroxide at 121 °C for 60 min, before soluble reactive phosphorous was determined using the molybdate-ascorbic acid method (Greenberg et al. 1985).

Fatty acid analyses

Fatty acids from supplemented and unsupplemented algae, from snail faecal pellets and from snail soft bodies were extracted with dichloromethane : methanol (2:1, vol/vol) as described by Von Elert and Stampfl (2000). PUFAs were quantified as fatty acid methyl esters (FAMEs) using a HP 6890 GC (Agilent Technologies, Waldbronn, Germany) equipped with a DB 225 fused silica column (J&W Scientific, Folsom, USA) and a flame ionisation detector with heptadecanoic acid methyl ester and tricosanoic acid methyl ester as internal standards as Von Elert and Stampfl (2000). Identification of the FAMEs was based on comparison of retention times to those of reference compounds.

Statistical analyses

R. ovata soft body growth rates (mg Ind^{-1} d^{-1}) were calculated and it was tested for differences between treatments using a one-way analysis of variance (ANOVA), followed by Tukey HSD post-hoc comparisons. Furthermore, linear correlation analyses (Pearson's R) were performed to investigate the hypothesised connection between the diet's content of total and n-3 PUFAs and *R. ovata* growth rates on the respective diets. All statistical analyses were performed using the STATISTICA v.6 software package (StatSoft 2004) and a significance level of α=0.05.

Results

Elemental composition of the algae

The algal food suspensions had carbon : nitrogen : phosphorous (C:N:P) ratios ranging from 34:4:1 in *Achnanthidium* sp. to 208:16:1 in *Z. circumcarinatum*. Thus, food C:N:P ratios were around the "Redfield-Ratio" considered optimal for the growth of benthic algae (Hillebrand and Sommer 1999), and were below the C:N:P ratios of *R. ovata* (Fink and Von Elert *submitted*). Therefore, the availability of the mineral nutrients N and P did most likely not limit the snails' growth on any of the algae offered.

Fatty acid composition of the algae

Total fatty acids in the food algae ranged from 124 µg mg C^{-1} for *G. parvulum* to 501 µg mg C^{-1} for *Z. circumcarinatum* (Tab. 1). There was no connection between algal taxonomic group and fatty acid content, neither for the total fatty acids, nor for the n-3 PUFAs, i. e. the diatoms did not contain consistently more or less of these fatty acids than the green algae analysed. *Achnanthidium* sp. contained high amounts of palmitoleic acid (16:1 n-7) and the highest observed concentrations of EPA. *G. parvulum* had the lowest amount of total and n-3 fatty acids of the algae analysed here. On the other hand, the major fatty acid in *G. parvulum* was the highly unsaturated eicosanoid EPA. *G. parvulum* did not contain any detectable amounts of α-linolenic acid (18:3 n-3, ALA), unless this PUFA was added artificially (Tab. 1). Supplementation of *G. parvulum* with ALA also led to a considerable increase in the sums of total and n-3 fatty acids. Similar to *G. parvulum*, the green alga *U. fimbriata* had comparatively low amounts of total fatty acids and n-3 PUFAs. The major fatty acid was ALA and no eicosanoids (such as EPA) nor any other fatty acids with a chain length larger than 18 carbons were detected in this alga. Supplementation of *U. fimbriata* with EPA resulted in EPA contents comparable to the natural amounts in the diatom *G. parvulum* (Tab. 1).

Table 1: Fatty acid content of *Achnanthidium* sp. (ACH), *G. parvulum* (GOM), *G. parvulum* + α-linolenic acid (GOM+ALA), *U. fimbriata* (ULO), *U. fimbriata* + eicosapentaenoic acid (ULO+EPA), and *Z. circumcarinatum* (ZYG). Amounts in bold are artificially added fatty acids not present in the original culture. Values are given as means ± SD of n = 3 for GOM and ULO, n = 2 for ACH, GOM+ALA, and ZYG and n = 1 for ULO+EPA. n.d. = not detected.

Fatty acid	ACH [µg mg C^{-1}]	GOM [µg mg C^{-1}]	GOM+ALA [µg mg C^{-1}]	ULO [µg mg C^{-1}]	ULO+EPA [µg mg C^{-1}]	ZYG [µg mg C^{-1}]
14:0	14.8 ± 0.8	n.d.	3.8 ± 5.4	n.d.	n.d.	7.1 ± 0.2
15:0	2.4 ± 0.0	n.d.	n.d.	n.d.	n.d.	n.d.
16:0	70.5 ± 0.4	25.8 ± 2.9	72.4 ± 21.2	19.0 ± 1.2	27.4	101.0 ± 0.7
16:1 n-7	223.4 ± 3.4	27.8 ± 5.4	103.0 ± 57.8	n.d.	n.d.	n.d.
18:0	4.2 ± 0.1	9.3 ± 0.9	10.2 ± 2.4	1.1 ± 1.8	12.2	9.9 ± 0.1
18:1 n-7	3.0 ± 0.0	n.d.	11.7 ± 0.8	3.1 ± 0.2	n.d.	102.9 ± 5.7
18:1 n-9/n-12	5.0 ± 0.1	3.4 ± 2.9	n.d.	1.0 ± 1.8	n.d.	5.1 ± 0.1
18:2 n-6	n.d.	n.d.	4.7 ± 0.3	9.3 ± 0.8	14.5	127.2 ± 7.5
18:3 n-6	2.7 ± 0.0	n.d.	n.d.	n.d.	n.d.	12.6 ± 0.7
18:3 n-3	3.1 ± 0.1	n.d.	**80.1** ± 6.4	91.0 ± 7.7	117.0	115.9 ± 5.9
18:4 n-3	4.5 ± 0.1	n.d.	n.d.	4.9 ± 0.3	n.d.	11.0 ± 0.6
20:3 n-6	6.3 ± 0.1	n.d.	n.d.	n.d.	n.d.	n.d.
20:4 n-6	n.d.	11.8 ± 0.7	28.0 ± 5.5	n.d.	n.d.	n.d.
20:5 n-3	60.2 ± 1.1	45.6 ± 2.7	58.6 ± 11.8	n.d.	**44.9**	8.4 ± 1.0
∑ fatty acids	400.2	123.6	372.4	129.3	216.0	501.0
∑ n-3	67.8	45.6	138.7	95.8	161.9	135.2

The other green algal strain used in this study, *Z. circumcarinatum*, contained the highest amounts of total fatty acids in all the analysed taxa. Among the major fatty acids in *Z. circumcarinatum* were palmitic acid (16:0), vaccenic acid (18:1 n-7), linoleic acid (18:2 n-6) and ALA (18:3 n-3). Only traces of EPA were found in *Z. circumcarinatum* (Tab. 1).

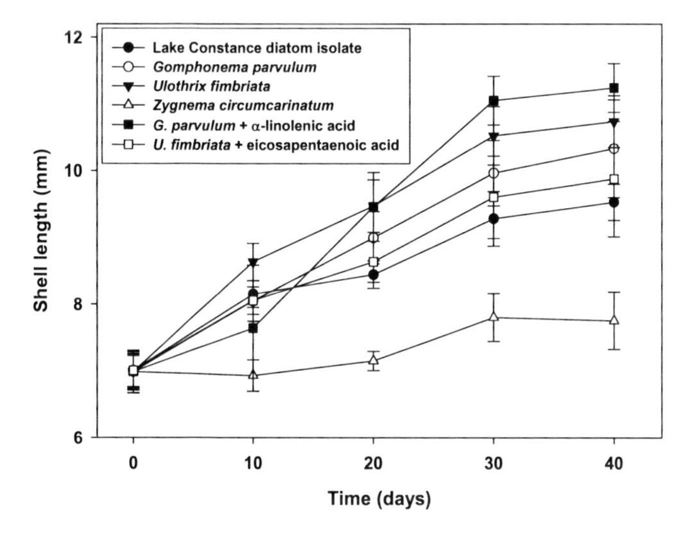

Figure 1: Shell growth of juvenile *R. ovata* on six different diets. Every 10 d, shell lengths of the experimental animals were determined. The diatom *G. parvulum* and the green alga *U. fimbriata* were fed to the snails both with their natural fatty acid composition and after addition of a single polyunsaturated fatty acid. Values are means ± SD of n=5 replicates.

R. ovata growth

All juvenile *R. ovata* fed one of the six algal suspensions ingested the offered food and snail growth was observed in all treatments over the course of the experiment (Fig. 1, 2). However, there were large differences in the growth of the snails between the different food sources. Snails which fed on the green alga *Z. circumcarinatum* showed the lowest growth in both shell length (Fig. 1) and SBDW (Fig. 2). Results of an ANOVA showed highly significant effects of the food treatments on the soft body growth rates of *R. ovata* ($F_{(5, 23)}$=22.7, p<0.001). A Tukey's post-hoc comparison revealed that growth on *G. parvulum*, *G. parvulum* + ALA and *U. fimbriata* was highest (Fig. 2). However, there was neither a positive correlation between the sum of total fatty acids (R = –0.61) and the snails' growth rates, nor between the sum of n-3 PUFAs (R = –0.14) and *R. ovata* growth rates, as indicated by correlation analyses. Despite having the highest fatty acid content of all the algae analysed here, *Z. circumcarinatum* was of significantly lower food quality than the other food types, as it supported only growth rates of 0.063 mg ind^{-1} d^{-1} when offered at the same food quantity (Fig. 2). Neither the addition of ALA to *G. parvulum*, nor the supplementation of *U. fimbriata* with EPA had significant effects on the snails' soft body growth rate as compared to the snails feds with the respective unmodified alga (Fig. 2).

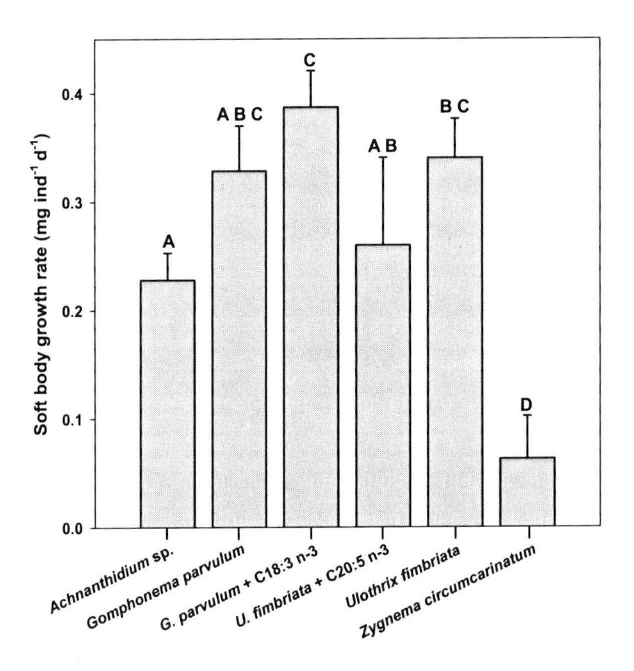

Figure 2: Soft body growth rates of R. ovata in the growth experiment. Growth rates were calculated as incremental increase in soft body dry weight per individual and day. The diatom *G. parvulum* and the green alga *U. fimbriata* were fed to the snails both with their natural fatty acid composition and after addition of a single polyunsaturated fatty acid. Values are means ± SD of n=5 replicates. Bars labelled with the same letters are not significantly different (Tukey's HSD following ANOVA).

Fatty acid composition of R. ovata soft bodies

When juvenile *R. ovata* were fed algae modified in their PUFA content for 41 days, their soft bodies' PUFA composition changed likewise (Fig. 3). When the snails were fed the diatom *G. parvulum*, which contains no ALA under natural conditions, only traces of ALA were found in the snails' soft bodies at the end of the experiment.

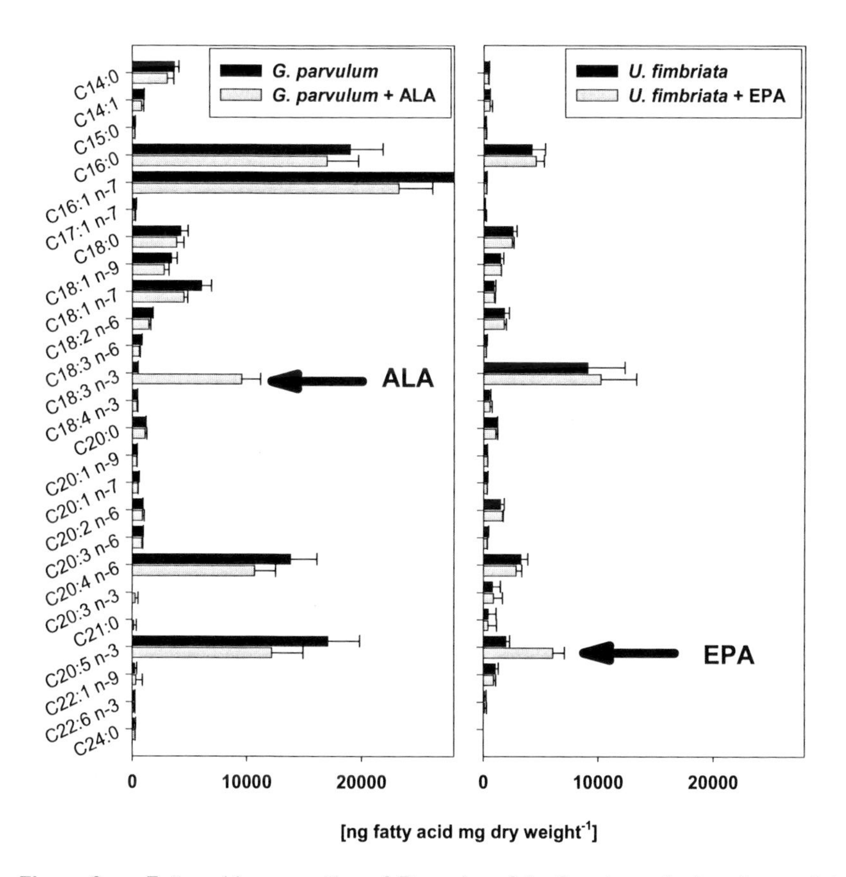

Figure 3: Fatty acid composition of *R. ovata* soft bodies dependent on the snails' diet. The diatom *G. parvulum* and the green alga *U. fimbriata* were fed to the snails both with their natural fatty acid composition and after addition of a single polyunsaturated fatty acid. α-Linolenic acid (18:3 n-3, ALA) was added to *G. parvulum*, and eicosapentaenoic acid (20:5 n-3, EPA) was added to *U. fimbriata*. Fatty acid content of the snails was determined after the experiment by extraction of total lipids and a subsequent analysis of fatty acid methyl esters in the transesterified lipid extract. Values are means ± SD of n=3 separately extracted snails per treatment. Arrows indicate PUFAs that had been artificially added to the snails' diet.

Table 2: Fatty acid content of faecal pellets excreted by juvenile *R. ovata* fed *Achnanthidium* sp. (ACH), *G. parvulum* (GOM), *G. parvulum* + α-linolenic acid (GOM+ALA), *U. fimbriata* (ULO), *U. fimbriata* + eicosapentaenoic acid (ULO+EPA), and *Z. circumcarinatum* (ZYG). Values are given as means ± SD of n = 3. n.d. = not detected.

Fatty acid	ACH [µg mg dw^{-1}]	GOM [µg mg dw^{-1}]	GOM+ALA [µg mg dw^{-1}]	ULO [µg mg dw^{-1}]	ULO+EPA [µg mg dw^{-1}]	ZYG [µg mg dw^{-1}]
14:0	0.3 ± 0.1	0.1 ± 0.1	0.2 ± 0.1	0.1 ± 0.1	n.d.	0.1 ± 0.2
16:0	1.8 ± 0.6	1.3 ± 0.6	1.3 ± 0.8	0.3 ± 0.1	0.3 ± 0.1	1.9 ± 1.3
16:1 n-7	5.3 ± 1.8	1.3 ± 0.7	1.2 ± 0.7	0.1 ± 0.0	n.d.	n.d.
17:1 n-7	0.1 ± 0.0	n.d.	n.d.	n.d.	n.d.	n.d.
18:0	0.2 ± 0.0	0.1 ± 0.1	0.2 ± 0.1	0.1 ± 0.1	0.1 ± 0.1	0.1 ± 0.2
18:1 n-9/n-12	0.1 ± 0.1	n.d.	0.2 ± 0.1	0.1 ± 0.1	n.d.	0.3 ± 0.4
18:1 n-7	n.d.	0.1 ± 0.1	n.d.	n.d.	n.d.	0.6 ± 0.6
18:2 n-6	0.2 ± 0.1	n.d.	n.d.	n.d.	n.d.	1.6 ± 1.0
18:3 n-3	n.d.	n.d.	0.2 ± 0.3	n.d.	0.1 ± 0.1	2.9 ± 0.6
18:4 n-3	0.1 ± 0.1	n.d.	n.d.	n.d.	n.d.	0.6 ± 0.4
20:4 n-6	0.5 ± 0.2	0.2 ± 0.3	0.2 ± 0.1	n.d.	n.d.	n.d.
20:5 n-3	1.7 ± 1.5	n.d.	n.d.	n.d.	n.d.	n.d.
24:0	0.1 ± 0.1	n.d.	n.d.	n.d.	n.d.	n.d.
∑ fatty acids	10.2	3.1	3.5	0.5	0.4	8.0
∑ n-3	1.7	n.d.	0.2	n.d.	0.1	3.5

However, juvenile *R. ovata* fed an ALA-enriched culture of *G. parvulum* for 41 days, considerable amounts (9.6 µg mg $SBDW^{-1}$) of ALA were detected in the soft bodies of the gastropods (Fig. 3). The green alga *U. fimbriata* does not contain any long chain PUFAs with chain length larger than C_{18}, and thus, no EPA (Tab. 1). *R. ovata* always contained relatively high amounts of EPA, irrespective if the snails' diet had contained EPA or not. However, the EPA content of the soft bodies of *R. ovata* fed on *U. fimbriata* which were artificially supplemented with EPA, was more than three times as high (6.1 µg mg $SBDW^{-1}$) than the EPA content of *R. ovata* fed the EPA-free *U. fimbriata* (Fig. 3).

Fatty acid composition of R. ovata faecal pellets

The fatty acid content of all analysed faecal pellets was very low (Tab. 2) compared to the algae that were fed to the snails. Only between 0.4 and 10.2 µg fatty acids were found per mg faecal dry weight. Only traces of PUFAs were found in *R. ovata* faecal pellets. The faeces' content of the n-3 PUFAs ALA, EPA and stearidonic acid (18:4 n-3) were often below the detection limit (Tab. 2). Neither snails fed the EPA-free *U. fimbriata* nor animals fed the EPA-enriched *U. fimbriata* suspension released any detectable amounts of EPA into their faecal pellets. Similarly, neither in the faeces of snails fed unmodified *G. parvulum*, nor in the faeces of *R. ovata* fed *G. parvulum* supplemented with ALA, any ALA could be detected. Thus, supplementation of dietary algae with single PUFAs did influence the PUFA content of both the algae and the snails that had been feeding on this PUFA-enriched alga. However, this PUFA supplementation did not influence the PUFA content of the snails' faeces.

Discussion

The different species of diatoms and green algae were of significantly different nutritional value for the growth of *R. ovata*, indicating that pronounced food quality differences do exist. However, there was no obvious connection between the algae's fatty acid content and the resulting growth increment of *R. ovata* fed these algae. On the contrary, the algae with the highest fatty acid content, *Achnanthidium* sp. and *Z. circumcarinatum* were of significantly inferior quality for the growth increment of *R. ovata* than *G. parvulum* and *U. fimbriata*, which had lower total fatty acid contents.

Furthermore, neither the addition of ALA to an ALA-free diatom, nor the supplementation of EPA to a green alga that lacks C_{20} PUFAs significantly affected the growth rates of juvenile *R. ovata*. Hence, in contrast to the results obtained with *Daphnia* (Wacker and von Elert 2001) or *D. polymorpha* (Wacker and von Elert 2002), no experimental evidence could be provided that PUFAs in general, and α-LA and EPA in particular, are limiting resources for the growth of this freshwater gastropod.

Especially the food's content of EPA appears to be of minor importance for the growth of *R. ovata*. This is surprising, since in all *R. ovata* analysed, EPA was among the major fatty acids detected, irrespective of the diet's EPA content. Even snails which had been fed an EPA-free diet still contained considerable amounts of EPA in their soft body tissues. This indicates, that *R. ovata* has the ability for the conversion of precursor molecules into EPA, so that sufficient amounts of EPA for the metabolic needs of this gastropod can be synthesised independent from the availability of EPA in the snails' diet. Similar to EPA, also ALA was found in all samples of *R. ovata* soft bodies, even if the snails had been fed exclusively with the ALA-free diatom *G. parvulum* for more than 40 days. *Daphnia* are known to be able to convert dietary ALA into EPA by enzymatic elongation and desaturation (Von Elert 2002). Probably, *R. ovata* is not only able to convert ALA into EPA, but also the other way round. This could allow the snails to partially compensate the absence of particular long chain PUFAs and be an important mechanism for these herbivores to cope with fluctuating contents of their periphyton resource. This ability for PUFA conversion in *R. ovata* can probably explain, why varying PUFA availabilities did not result in significant effects on the growth of juvenile *R. ovata*. So far, it was not tested, if the reproduction of these gastropod molluscs could be influenced by the EPA content of the food as it has been demonstrated for the freshwater bivalve *D. polymorpha* (Wacker and von Elert 2003, Wacker and von Elert 2004).

Furthermore, an effect of ALA-supplementation to *G. parvulum* on the growth of juvenile snails was indicated by the data, but was not statistically significant. This might be due to the high variation between snails in the experimental treatments, resulting partially from the unknown previous conditions of the field-collected animals. However, in contrast to growth experiments with juvenile *D. polymorpha* (Wacker and von Elert 2002), where the variance between replicates (n = 3) was extremely low, it was not possible to reduce the variance in the snail growth experiments despite a higher number (n=5) of replicates. The main reason is probably that genetically determined differences even between siblings originating from the same clutch are very high in freshwater snails (P. Fink, pers. observation) and hence, a reduction of the variance innate to the system is particularly difficult with these organisms. It cannot be excluded that further experiments would reveal a limitation of *R. ovata* growth by the availability of ALA and thus, the importance of PUFAs for the performance of this freshwater gastropod is not yet fully clear.

Interestingly, the results presented here for *R. ovata* are similar to those found for the prosobranch snail *Bithynia tentaculata* (Fink and Von Elert *in press*). Also for *B. tentaculata*, no experimental evidence for a growth constraint by the lack of PUFAs could be obtained. However, as the aforementioned study was hampered by some experimental difficulties, the present study was undertaken to investigate the role of PUFAs for the growth of freshwater snails in more detail. Nevertheless, as the observed effects are remarkably similar even for two only very distantly related gastropod species, the lack of a PUFA-limitation for snail growth is probably a finding applicable to herbivorous gastropods in general.

Only traces of PUFAs were found in faecal pellets excreted by *R. ovata* in the experiment. There are various possible reasons for the low amounts of PUFAs found in the faecal pellets. One reason could be a very efficient uptake mechanism during the snails' digestive process. Such an effective mechanism could enable the snails to use even traces of dietary PUFAs for their metabolism and would probably be highly adaptive if PUFAs are an important factor in snail nutrition. However, there could be other reasons than digestive uptake for the low amounts of PUFAs detected in the faecal pellets. First, PUFAs could have been leached from the faecal pellets between excretion and sampling, because sampling of faecal pellets was done only every 48 hours. Especially the highly unsaturated PUFAs are more hydrophilic than the saturated fatty acids and therefore, leaching of PUFAs might be relatively high in comparison to saturated fatty acids. Second, abiotic and biotic degradation processes probably reduced the amount of PUFAs within the faecal pellets. Polyunsaturated fatty acids such as EPA are highly susceptible to oxidative degradation. Furthermore, biogenic PUFA degradation by associated microorganisms is likely to have played a role.

It is interesting to observe that the algae supporting the lowest growth of *R. ovata* (*Achnanthidium* sp. and *Z. circumcarinatum*) led to the highest concentrations of both total fatty acids and n-3 PUFAs in the snails' faecal pellets when the gastropods fed on these algae. Probably these comparably high amounts of fatty acids resulted from uningested pseudofaeces produced by *R. ovata* when fed either *Achnanthidium* sp. or *Z. circumcarinatum* rather than from ‚true' faecal material that had passed the snails' digestive tract. This could indicate a complete removal of PUFAs during R. ovata's digestive process. However, it remains to be tested, whether this removal is due to active uptake by the snails' digestive apparatus or due to a complete degradation of PUFAs due to gut-associated microorganisms.

The most probable reason for the finding that the nutritional quality of the green alga *Z. circumcarinatum* for *R. ovata* was much lower than for the other three algal taxa lies in the stiff and rigid filaments of this benthic chlorophycean. Possibly, these filaments could not be ingested with equal efficiency as the other food organisms. And even when ingested, the rigid cell wall of Z. circumcarinatum filaments might have at least partially resisted digestion by the snails. A similar mechanism of resistance to digestion has been found in the unicellular planktonic green alga *Scenedesmus* (Van Donk and Hessen 1993).

In conclusion, the morphology of algal filaments (*Z. circumcarinatum*) and the algal cells' content of nitrogen and phosphorous (Fink and Von Elert submitted) are probably much more important in determining algal food quality for *R. ovata* than the dietary content of PUFAs. The cause for the snails' low susceptibility for limitation by the diet's PUFA content is probably the gastropods' ability for conversion of dietary PUFAs into each other. This would allow these herbivores to cope with fluctuating availabilities of PUFAs in their diet and might be a factor in explaining the tremendous ecological success of this group of organisms in littoral zone ecosystems.

Acknowledgements

We thank O. Walenciak and C. Gebauer for analyses of algal carbon, nitrogen and phosphorous and A. Fibich for experimental assistance. Helpful comments by L. Peters greatly improved an earlier draft of this manuscript. This study was supported by the Deutsche Forschungsgemeinschaft (DFG) within the Collaborative Research Centre SFB 454 – „Littoral of Lake Constance".

Chapter 6

Volatile foraging kairomones in the littoral: attractance of algal odors for a herbivorous freshwater gastropod

Patrick Fink, Eric Von Elert and Friedrich Jüttner

Abstract

Volatile organic compounds (VOCs) produced by photoautotrophic organisms are often considered to be the primary source of foul source-water odors. However, their biological functions often remain obscure. Recently, research focused on volatile aldehydes released upon grazing by marine diatoms that have toxic effects on the reproductive capacity of herbivorous crustaceans and thus are considered to form an activated antipredator defense of these algae. But VOCs may also serve as important infochemicals, especially in biofilms of benthic algae and cyanobacteria. Benthic mat-forming cyanobacteria have been shown to produce VOCs that can be used as habitat-finding cues by insects, nematodes and possibly by other organisms. Whenever investigations on food preference are performed, a clear distinction between odorous infochemicals working over distance and taste (which requires direct contact with the food source) has to be made. Here, with the aid of a newly developed behavioral assay for gastropods that allows detection of snail food preference without offering food, and thus making possible the distinction between taste and odor in food preference, we demonstrate that VOCs released from a benthic, mat-forming green alga (*Ulothrix fimbriata*) are attractive to a herbivorous freshwater snail (*Radix ovata*). The volatiles released from the algae can be utilized as food-finding cues ('foraging kairomones') by the herbivorous snails. While a mixture of compounds was clearly attractive to the snails, single fractions were no longer attractive. This indicates that a bouquet of volatiles, rather than single compounds, is responsible for the attractant activity. This study shows that volatile organic compounds, until now seen primarily as important information transmitting cues in terrestrial ecosystems, can also play a steering role as infochemicals in freshwater benthic habitats.

Introduction

Aquatic systems are ideally suited for communication via chemical cues, as infochemicals can be easily distributed in concentrations sufficient for response (Wisenden 2000). Many groups of aquatic organisms have developed the ability to optimize their fitness based on the perception of infochemicals related to predation (Von Elert and Loose 1996), reproduction (Müller et al. 1971) and foraging (Thomas et al. 1980). Infochemicals should, in general, be expected to be important in benthic systems with their highly diverse physical structure and in particular for animals with a poorly developed visual sense (Wisenden 2000).

Volatile organic compounds (VOCs) produced by algae and cyanobacteria are a frequent nuisance in water treatment, especially for drinking water. Musty and earthy smells, frequently attributed to cyanobacterial blooms cause high economical costs as the removal of these odor compounds is difficult and expensive (Watson and Ridal 2004). On the other hand, the biological function of these odor compounds is largely unknown (Watson 2003). Recently, research focused on volatile aldehydes that could play a role in an activated defense mechanism of marine diatoms (Pohnert 2000). α, β, γ, δ-Unsaturated aldehydes (Pohnert et al. 2002) released from wound-activated diatom cells drastically lowered the hatching rate of copepod eggs and might therefore serve as defense on the population level for the diatoms (Miralto et al. 1999, Ianora et al. 2004).

Watson (2003) hypothesized an alternative role of VOCs rather than in activated defenses. The chemical properties of VOCs make them well suited as infochemicals in aquatic ecosystems. And indeed, VOCs have been described to function as habitat-finding cues for both aquatic insects (Evans 1982) and nematodes

(Höckelmann et al. 2004). In particular, VOCs seem to be suitable as infochemicals in the highly structured benthic, rather than in planktonic systems, as they remain more localized and are not diluted too quickly (Watson 2003).
Recently, mats of benthic diatoms have been shown to release various VOCs upon cell lysis (Jüttner and Dürst 1997), and benthic cyanobacterial assemblages have been identified as a major source of the volatile nuisance compounds geosmin and 2-methylisoborneol (Watson and Ridal 2004). Thus, the ecological function of VOCs produced by benthic primary producers probably merits further attention (Watson 2003).
In littoral ecosystems, both marine and freshwater, herbivorous gastropods are of major importance, as they structure the algal community by strong top-down predation pressure, and also form an important link to higher trophic levels, as they are preyed upon by both fish and invertebrate predators such as crayfish (Turner et al. 2000). Gastropods generally have rather low visual capabilities that they could make use of for the localization of food patches (Gal et al. 2004), and their locomotion is associated with considerable energetic costs, primarily due to the production of pedal mucus (Denny 1980). Therefore, it should be highly adaptive for snails to rely on chemical cues in order to minimize these costs by directed chemotaxis towards potential food sources.
It is well known that most of the behavioral repertoire of gastropods is affected by chemical stimuli and that, in particular, gastropods use their sensitivity to chemical cues as a principal modality for the detection of distant objects in the environment (Croll 1983). So far, food preference of gastropods in choice experiments was quantified only either as residence time on a particular food patch or as the amount of food consumed (e.g., Madsen 1992, Brendelberger 1995b, Wakefield and Murray 1998) and it was not investigated, if long-range chemical cues were involved in the observed food preferences. On the other hand, snail chemotaxis towards dissolved sugars, amino acids and carboxylic acids was determined only in the context of control mechanisms for snails that are intermediate hosts for parasites (Thomas et al. 1980, Thomas 1986). However, it remains unclear how these dissolved cues are related to the process of food-finding. Therefore, the interaction between food preference and attractance by infochemicals remains unresolved.
VOCs released from benthic algae are both suitable for dispersion over distance and a good predictor of the presence of algal food. This led us to the hypothesis, that

freshwater gastropods could utilize algal volatiles as foraging kairomones (sensu Ruther et al. 2002) for food-finding.

To experimentally investigate this hypothesis, we first determined the bouquet of VOCs released upon cell lysis from the filamentous benthic green alga *Ulothrix fimbriata* (BOLD), a common member of periphyton communities in fresh waters. Further, in standardized assay systems, we investigated the behavioral response of the common pulmonate snail *Radix ovata* (DRAPARNAUD) to this bouquet of VOCs. Different mixtures of the identified VOCs were assayed with the aim to identify the compounds responsible for this attractivity.

Methods

Cultures

An axenic culture of the filamentous benthic green alga *Ulothrix fimbriata* (BOLD) was obtained from the Sammlung für Algenkulturen, Göttingen, Germany (SAG 36.86) and cultivated in suspension cultures semicontinously at a dilution rate of 0.25 d^{-1} in WC medium (Guillard and Lorenzen 1972) at a constant temperature of 20° C and a light intensity of $1 \cdot 10^{16}$ quanta $\cdot$ s^{-1} cm^{-2}. Heavy bubbling of the culture with sterile, compressed air was applied to prevent sinking and attachment of the benthic algae to the bottom of the culture flask. Algae were harvested, concentrated by centrifugation at 4000 x g and resuspended in 0.45 µm filtered Lake Constance water. In the resulting food suspensions, carbon concentrations were adjusted to 0.5 mg particulate organic carbon per ml using the photometric light extinction at 800 nm and carbon-extinction equations previously determined.

Juvenile *Radix ovata* (DRAPARNAUD) with shell lengths ranging from 5-10 mm were collected in the littoral zone of Lake Constance and acclimatized in the laboratory (at 20° C and constant dim light) prior to use in the choice experiments. During the acclimation period, snails were fed Tetra PlecoMin® fish food pellets (Tetra, Melle, Germany). Prior to the experiments, the animals were kept for 24 hours in 0.45 µm filtered Lake Constance water without food to increase their food searching activity due to moderate starvation.

VOC analyses

Algal biomass equivalent to 10 mg particulate organic carbon was lysed by freeze-thawing prior to closed-loop stripping in 40 ml ultrapure water with 25 % NaCl p.a. onto Tenax TA adsorbent (Supelco) as described by Jüttner (1988b). Subsequently, the volatiles were thermally desorbed from the Tenax material and transferred directly onto a capillary GC/MS column (DB 1301 J&W Scientific, Folsom, CA, USA, 30 m length and 0.32 mm i. d.). Helium was used as transfer and carrier gas. Gas chromatography-mass spectrometry (GC/MS; Thermo/Finnigan GCQ) was applied to identify volatiles produced by *U. fimbriata*. The applied GC temperature program was 4 min at 0° C and 5° C min^{-1} to 250° C, followed by another 10 min at 250° C. Volatile organic compounds were identified by comparison of retention times and mass fragmentation patterns (EI at 70 eV) with reference compounds (Aldrich).

The main volatiles identified in *U. fimbriata* were quantified as described by Jüttner (1988b) using the peak areas of characteristic mass fragments and a calibration curve for each compound with 3-hexanone as internal standard. 3-Hexanone, which does not occur in *U. fimbriata*, was chosen as a reference substance of intermediate volatility with respect to the identified substances.

VOC extraction for choice assays

Volatiles were extracted from the aqueous phase via closed loop stripping onto the adsorbent Tenax TA as described above. Then, the adsorbent was eluted with 5 ml diethyl ether. The ether was gently evaporated to dryness with nitrogen and the residue immediately taken up in 100 µl of ethanol. This ethanolic extract was stored in gas-tight vials at -20° C until use in the choice assay, but never longer than 48 hours to avoid loss of highly volatile compounds. Prior to the choice assay, the ethanolic extract was diluted in 7 ml 0.45 µm filtered Lake Constance water, and this aqueous solution was added into the containers in the choice aquarium (see below).

As control treatment, 40 ml of ultrapure water were stripped with 25 % NaCl and eluted with diethyl ether as described above. To exclude possible effects of contaminants introduced during the cultivation of the algae, another control series of experiments was performed in which WC medium (aerated for several days like the algal culture) was used instead of the ultrapure water.

Synthetic VOC mixtures

Synthetic mixtures of pure compounds (Aldrich) of VOCs identified in *U. fimbriata* were formulated to approximately match the concentrations in the algal VOC extract (Tab. 1). However the (*Z*) isomer of 2-pentenal and the (*E,Z*) isomer of 2,4-heptadienal were not available, thus only the (*E*) and (*E,E*) isomers respectively, were used for the complete VOC-mixture (Tab. 1). In order to resolve whether the attractant activity of *U. fimbriata* volatiles was dependent on a specific class of substances or rather on a multicomponent odor, two additional mixtures of VOC reference compounds were produced (Tab. 1); one contained only the three C_5 compounds 1-penten-3-one, 1-penten-3-ol, and (*E*) 2-pentenal (C_5-mix), the other only the C_7 compound (*E,E*) 2,4-heptadienal (C_7-mix).

Table 1: Composition of the synthetic VOC mixtures offered to *Radix ovata* in the food choice assays; the complete VOC-mix was designed to mimic the VOC bouquet of *U. fimbriata*, the C_5-Mix contained only the C_5 compounds present in the complete VOC-mix (1-penten-3-one, 1-penten-3-ol and (*E*)2-pentenal), and the C_7-Mix contained only the C_7 compound (*E, E*) 2,4-heptadienal present in the complete VOC-mix; amounts given were added to the containers in the choice assays.

VOC	Supplier and product no.	(µg per container in the choice assay)		
		Complete mix	C_5-mix	C_7-mix
1-Penten-3-one	Aldrich E5,130-9	84.5	84.5	--
1-Penten-3-ol	Aldrich P860-2	83.9	83.9	--
(*E*) 2-Pentenal	Aldrich 26,925-5	86.0	86.0	--
(*E, E*) 2,4-Heptadienal	Aldrich 18,054-8	0.88	--	0.88

Setup of the choice assays

The setup of the choice assay was specifically designed to separate chemotaxis, mediated through foraging kairomones, from pure food preference resulting from taste-receptor mediated effects on patch residence time. Therefore, it was important to offer the food source in a way that allowed for the release and detection of foraging kairomones, but prevented the experimental animals from accessing the food itself.

The choice assays were performed in an aquarium (320 x 170 mm and 180 mm deep, total volume 10 L), which was placed in a climate controlled room at 20° C and filled with 1 L of 0.45 µm filtered Lake Constance water. The setup of the choice assays was developed to be suitable for whole algal cells as well as for extracts of volatiles (diluted with water). Special containers were designed to allow for introduction of samples without disturbing the water body (Fig. 1A, B). They were modified from the „olfactometers" described by (Thomas et al. 1980) and consisted of two cylindrical perspex rings with radial bores (5 mm diameter) near the bottom side of the ring. The inner ring was closed at the bottom side by gluing of a circular perspex plate to one opening and had a diameter (40 mm) that fitted exactly into the outer ring (Fig. 1a, b). Hence, by rotation of the outer ring it was possible to open the container by bringing the bores of the outer and inner ring to match (Fig. 1A). Likewise, the container could be closed by rotating the outer ring so that the bores did not match any more, which closed the container and stopped exchange of substances between inner and outer side of the container (Fig. 1B).

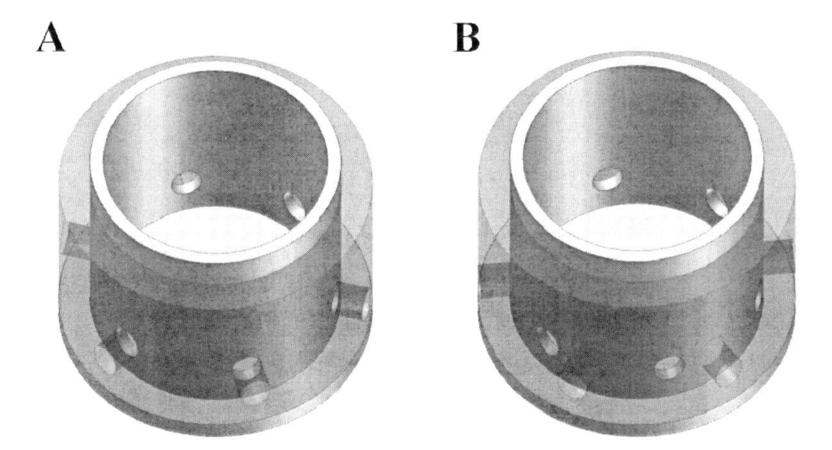

Figure 1: Schematic drawing of the containers for food choice assays. A) open position; B) closed position. For the assays two such containers were placed at the far ends of an aquarium (21 cm distance).

To initiate the assay, two closed containers were placed at the far ends of the aquarium (21 cm min. distance) and 7 ml of the respective sample (diluted with filtered lake water of the same temperature) were added into each of the containers so that the inner and outer water levels matched each other. The setup was designed to reach a water level that was higher than the radial bores of the containers, but lower than the opening at the top of the containers, which was covered using a circular glass slide. Then, containers were opened by a rotation of the outer rings to connect the inner and outer water bodies via the radial bores of the containers, and five juvenile *R. ovata* were introduced at the center of the aquarium. This marked the start of the experiment. From there on, the relative distance of each *R. ovata* individual to both containers was recorded every minute with a resolution of ± 1 cm for a total time period of 40 minutes. Therefore, the initial value for each snail at the beginning of the experiment (equal distance to both containers) was zero, and changed continuously during the experiment as the snails moved through the aquarium.

After opening the containers with the samples, dissolved substances could leave the containers through the bores. This was verified by the addition of 1-penten-3-one to one of the two containers. 30 minutes after opening of the container, a quantitative analysis of the aquarium water body (1 L) revealed that 16.6 % of the VOC in the container (1.7 $\mu g\ L^{-1}$) were released to the surrounding water body of the aquarium within half an hour.

Statistical analyses

As the five individual snails in each experiment were not independent from each other, the mean of their distribution was calculated for every reading at intervals of 1 minute. These mean values were plotted versus reading time and treated as one replicate experiment. All experiments were replicated at least five times, resulting in experimental series with n = 5-10 for each treatment that was to be tested. Between replicate experiments, the sides for the containers with the treatment and the control were exchanged to exclude directional effects introduced by the experimental setup.

First, a series of control experiments (n=14) was performed in which both containers were filled with filtered lake water. The distribution of the snails in this „control" series was tested against the series of experiments with treatment samples in one of the containers (and a control sample in the other container) by comparing experimental series with and without treatment samples via repeated-measurement analysis of variance (RM-ANOVA) using the GLM module of STATISTICA v.6 software package (StatSoft 2004) and a significance level of α=0.05.

Results

Analysis of VOCs from U. fimbriata

While undamaged *U. fimbriata* cells did not release any VOCs to the surrounding water, freeze-thawing of *U. fimbriata* led to a pronounced release of a variety of volatiles (Tab. 2). Among these VOCs, the most dominant group were the lipoxygenase products released upon cleavage of polyunsaturated fatty acids such as 1-penten-3-one, 1-penten-3-ol, (*Z*)2-pentenal, (*E*)2-pentenal, (*E,Z*)2,4-heptadienal and (*E,E*)2,4-heptadienal. As minor compounds, also some nor-carotenoids (6-methyl-5-hepten-2-one, α-ionone, β-ionone and β-cyclocitral), that are produced by enzymatic carotenoid degradation, occurred (Tab. 2).

Table 2: Volatile organic substances (VOCs) liberated from *Ulothrix fimbriata* (Chlorophyceae) upon freeze-thawing of cells; compounds are given with their molecular mass (M), grouped by their biosynthetic pathways and sorted by retention time (Rt).

Compound group	VOC	M (g mol^{-1})	Rt (min)
Fatty acid pathway	1-Penten-3-one	84	10.03
	1-Penten-3-ol	86	10.71
	(*Z*) 2-Pentenal	84	12.44
	(*E*) 2-Pentenal	84	12.83
	(*E*) 2-Hexenal	98	16.44
	(*E, Z*) 2,4-Heptadienal	110	21.62
	(*E, E*) 2,4-Heptadienal	110	22.06
	Nonanal	142	24.35
	Pentadecane	212	34.12
Isoprenoid pathway	6-Methyl-5-hepten-2-one	126	21.15
	β-Cyclocitral	152	27.35
	Geranylacetone	194	33.59
	α-Ionone	192	34.41
	β-Ionone	192	35.08

We determined the concentrations of the main compounds released by *U. fimbriata* upon cell lysis. Concentrations of VOCs are given in Table 3. The C_5 compounds 1-penten-3-one, 1-penten-3-ol, and (*Z*)2-pentenal were the most important compounds released by *U. fimbriata*, resulting in release of up to 0.9 µg 1-penten-3-ol per mg algal carbon. The (*E*) isomer of 2-pentenal and (*E,Z*) 2,4-heptadienal and (*E,E*) 2,4-heptadienal were of minor importance as their release only accounted for 10.4 - 18.7 % of the release of 1-penten-3-ol (Tab. 3).

Table 3: Quantities of the main VOCs released from *Ulothrix fimbriata*; compounds are sorted by retention time (Rt); values are given as ng substance per mg algal carbon; in the food choice assays, VOCs extracted from 10 mg algal carbon were added to each container.

VOC	Rt (min)	Release (ng mg POC^{-1})	Choice assay (µg per container)
1-Penten-3-one	10.03	535.9	5.4
1-Penten-3-ol	10.71	924.2	9.2
(*Z*) 2-Pentenal	12.44	487.7	4.9
(*E*) 2-Pentenal	12.83	172.8	1.7
(*E,Z*) 2,4-Heptadienal	21.62	169.9	1.7
(*E,E*) 2,4-Heptadienal	22.06	96.5	1.0

Food choice experiments with R. ovata

When a suspension of undamaged *U. fimbriata* cells was added to one of the containers (and filtered lake water to the other container), juvenile *R. ovata* did not show any chemotactic response, neither attractance nor repellence from the algae (Fig. 2A, Tab. 4), which indicates that they were incapable of responding to an infochemical gradient. Similarly, no response occurred when the supernatant of an exponentially growing *U. fimbriata* culture was offered (Tab. 4), which indicates that exudates of actively growing cells did not evoke a behavioral response. However, when the same biomass of *U. fimbriata* that was not attractive as intact cells was 'activated' (lysed) via freeze-thawing and when the VOCs that were released by this treatment were trapped on Tenax adsorbent, the diluted eluate of the Tenax adsorbent (containing only extracted volatiles) was clearly preferred by *R. ovata* over a control eluate (Fig. 2B, Tab. 4). This attractance was not introduced by the cultivation of the alga or by the VOC extraction process itself, as a VOC extract of aerated sterile culture medium without algae was not preferred over a control eluate

(Fig. 2C, Tab. 4). Hence, the attractivity clearly depended on the VOCs released from the algae. Similarly, a synthetic mixture of pure reference compounds designed to mimic the algal VOC bouquet (Tab. 1) was preferred over a solvent control by *R. ovata* (Fig. 2D, Tab. 4).

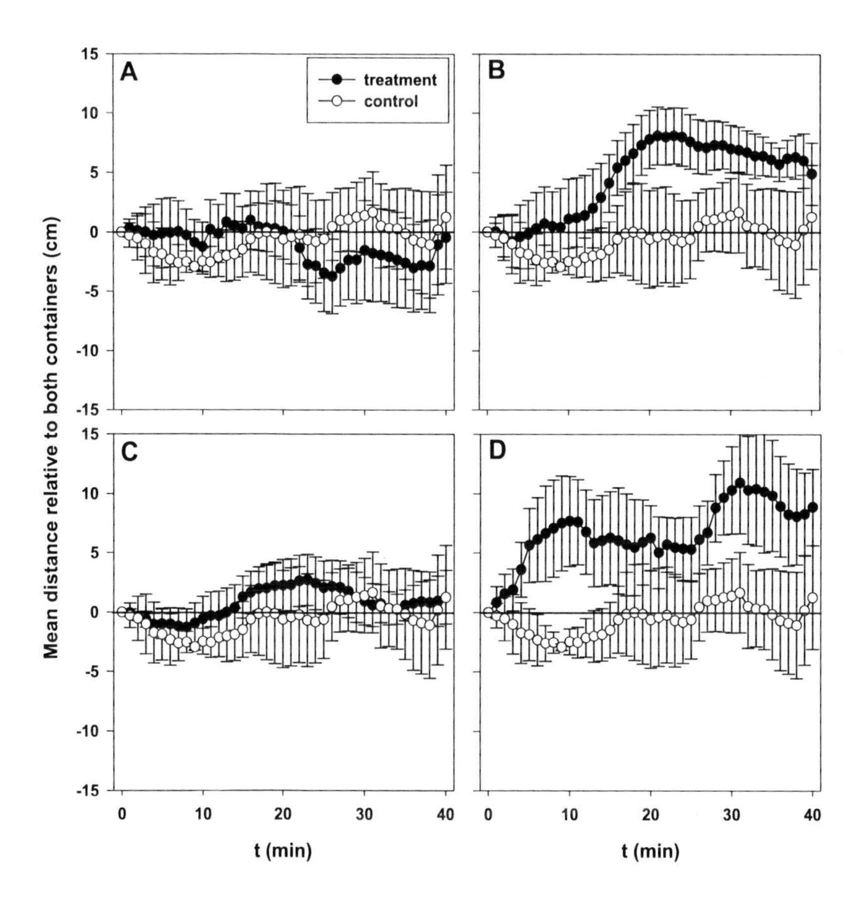

Figure 2: A container with a test extract was introduced into the aquarium at a distance of 10.5 cm from the center (scoring value +21) and a container with a control extract was introduced at the opposite end of the aquarium (scoring value -21). Depicted are the mean values (± SE) of the snails' relative distance to both containers in the choice assays. To the treatment containers the following test extracts were added (closed symbols): A) undamaged *U. fimbriata* cells (n=7); B) a VOC extract from lysed *U. fimbriata* cells (n=5); C) a VOC control extract from aerated algal medium (n=10); D) the synthetic complete VOC-mix containing C_5 and C_7 odor compounds (n=6). Results from a series of control experiments (both containers with filtered lake water) are plotted for comparison (open symbols).

Table 4: Results of repeated-measurement analyses of variances on the mean positions of 5 juvenile *R. ovata* in the food choice experiments. All analyses compared the treatment (choice between a given test treatment and a control) to a series of control experiments (n=14) in which snails had to choose between two containers with filtered lake water. A) choice between control and intact cells of *U. fimbriata* (n=7). B) choice between control and VOC extract of *U. fimbriata* (n=5). C) choice between control and VOC extract of aerated WC medium (n=11). D) choice between control and the complete VOC-mix (n=6). E) choice between control and the synthetic C_5-Mix (n=10). F) choice between control and the synthetic C_7-Mix (n=6). G) choice between control and culture supernatant from an *U. fimbriata* culture (n=5).

A)	SS	df	F	p	
Treatment	359.69	1	0.381	0.545	n.s.
Error	16972.49	19			
Time	395.20	39	0.418	0.999	n.s.
Time x Treat	1106.98	39	1.172	0.221	n.s.
Error	16998.87	702			

B)	SS	Df	F	P	
Treatment	3959.27	1	4.575	0.047	*
Error	14712.65	17			
Time	2019.12	39	3.205	0.000	***
Time x Treat	1234.34	39	1.959	0.001	***
Error	10710.56	663			

C)	SS	df	F	p	
Treatment	507.53	1	0.622	0.438	n.s.
Error	18752.72	23			
Time	843.75	39	0.992	0.486	n.s.
Time x Treat	491.19	39	0.577	0.983	n.s.
Error	19565.31	897			

D)	SS	Df	F	P	
Treatment	9239.39	1	8.501	0.009	**
Error	19564.60	18			
Time	1449.77	39	1.501	0.027	*
Time x Treat	956.60	39	0.991	0.488	n.s.
Error	17381.63	702			

E)	SS	Df	F	P	
Treatment	2715.27	1	2.185	0.154	n.s.
Error	27344.71	22			
Time	1876.39	39	1.434	0.043	*
Time x Treat	2608.89	39	1.994	0.000	***
Error	28768.75	858			

Table 4 *continued*

F)	SS	df	F	P	
Treatment	498.39	1	0.623	0.440	n.s.
Error	14397.08	18			
Time	753.09	39	0.804	0.799	n.s.
Time x Treat	1908.12	39	2.037	0.000	***
Error	16863.48	702			

G)	SS	df	F	P	
Treatment	338.56	1	0.411	0.530	n.s.
Error	14007.82	17			
Time	842.80	39	1.125	0.280	n.s.
Time x Treat	2341.34	39	3.125	0.000	***
Error	12738.88	663			

(asterisks indicate significant differences at p<0.05 (*), p<0.01 (**), and p<0.001 (***), n.s. = not significant)

In an attempt to distinguish if the attractivity was due to the C_5 or the C_7 compounds, two further sets of choice assays were performed in which either only a mixture of 1-penten-3-one, 1-penten-3-ol, and (*E*) 2-pentenal (C_5-Mix, Tab. 1), or (*E, E*) 2,4-heptadienal (C_7-Mix, Tab. 1) was offered to the snails. Both the C_5 and the C_7 fraction were not active when offered alone (Tab. 4). However, the mixture in which both compound groups were present, was clearly attractive to *R. ovata* (Fig. 2D, Tab. 4). This indicates that a bouquet of volatiles, rather than single compounds is responsible for the attractivity.

Discussion

VOCs released by U. fimbriata

Although numerous studies have investigated volatile compounds from diatoms (e.g., Pohnert and Boland 1996, Wendel and Jüttner 1996, Jüttner and Dürst 1997) and especially from cyanobacteria (Jüttner et al. 1983, Jüttner 1987, Watson and Ridal 2004), the ecologically very important group of green algae have so far been widely neglected with respect to their bouquet of odorous substances. Being related to higher plants it might be suspected that green algae exhibit similar patterns of volatiles. The so called „green leaf volatiles“ (GLVs) of terrestrial plants consist mainly of C_6 compounds (Jüttner 1987) and play an important role in herbivory-activated defense mechanisms (Kessler and Baldwin 2001, Halitschke et al. 2004).

The bouquet of VOCs released from the benthic green alga *U. fimbriata* upon cell damage is quite different from the bouquets of GLVs released by terrestrial plants (Tab. 2). In *U. fimbriata*, various lipoxygenase products are released. Such lipoxygenase products are reported to be released as cleavage products of fatty acids (Pohnert 2002). Among them are C_6 compounds such as (*E*)2-hexenal (and a not unambiguously identified and therefore not in Table 2 included (*Z*) 3-hexanal) that are also found in GLV mixtures from leaves; however, in contrast to higher plants, these seem to be not among the major compounds as the detected peaks were small in comparison with C_5 or C_7 compounds. In addition to VOCs resulting from fatty acid degradation, a variety of volatile nor-carotenoids are liberated that are probably degradation products of carotenoid photopigments (Jüttner 1988a, Simkin et al. 2004). Among others, β-ionone was released from *U. fimbriata*, which is an important component of flower (e.g. *Viola* sp.) odors, and was shown to be a repellent for the freshwater nematode *Bursilla monohystera* (Höckelmann et al. 2004).

Response of R. ovata to algal VOCs

While foraging kairomones can be sensed over a distance, taste as a mediator of food preference works only via direct contact with the food source. These mechanisms should be investigated separately. Our newly developed behavioral assay system allowed for a clear distinction between chemotaxis due to foraging kairomones and taste receptor mediated food preference. Volatiles liberated from *U. fimbriata* upon cell damage were significantly preferred by *R. ovata* over control extracts. Neither offering of undamaged algae, nor of a culture supernatant from an exponentially growing *U. fimbriata* culture, did lead to any detectable chemotactic response in *R. ovata*. Apparently, cell damage is necessary for the liberation of the infochemicals perceived by the snails. This is supported by the current ideas about the liberation of VOCs from algal phospholipids via a rapid enzymatic degradation (Pohnert 2002). This enzyme cascade is believed to start with a wound-activated phospholipase cleaving algal phospholipids and releasing free fatty acids. These free fatty acids, that can be potent toxins for benthic herbivores (Jüttner 2001), are in part rapidly oxygenated by a lipoxygenase that introduces dioxygen into the fatty acid molecule (Wendel and Jüttner 1996). Subsequently, the oxygenated fatty acid is cleaved by a specific lyase enzyme into a volatile compound and a non-volatile short-chain fatty acid (Pohnert 2002). Thus, wounding of algal cells seems to be an essential prerequisite for the formation of volatile oxylipins.

We could unequivocally demonstrate that VOCs and not other compounds released upon cell lysis were responsible for the attractant activity, as the eluate of the Tenax adsorbent loaded with *U. fimbriata* volatiles was clearly preferred over a control eluate. Only volatile organic compounds adsorb to Tenax TA, thus these compounds must have been responsible for the attractant activity of the eluate. This was further supported by the finding that a synthetic mixture of pure reference compounds designed to mimic the VOC bouquet of *U. fimbriata* was also highly attractive to *R. ovata*. To identify a specific size class of molecules responsible for the activity, we compared the attractant activity of only the C_5 and the C_7 fraction of the complete VOC mix. As neither the C_5- nor the C_7-fraction of the complete *U. fimbriata* VOC-bouquet was active alone, a multicomponent odor rather than a single substance seems to determine the attractivity for *R. ovata*. This is remarkably similar to the benthic freshwater nematode *Bursilla monohystera* (Höckelmann et al. 2004), which also did respond to a bouquet of cyanobacterial VOCs but not to single compounds. Also in terrestrial systems, multicomponent odors are frequently much more effective in eliciting responses in insects than single compounds (El-Sayed et al. 1999).

For the behavioral assays, we deliberately used snails that had not encountered odors from *U. fimbriata* before in order to focus exclusively on genetically fixed abilities for food location. However, snails are known to be able to learn to respond to stimuli they have encountered before (Croll 1983). Thus, snail food preferences are almost certainly influenced by both innate factors and olfactory learning (Croll and Chase 1980). Furthermore, freshwater gastropods are known to be able to adapt their digestive enzymes to optimally suit the digestion of the most abundant food source (Calow and Calow 1975, Brendelberger 1997b). Hence, the response in natural systems could be much higher, if olfactory learning and conditioning of the array of digestive enzymes plays a role, and thus, be a rather conservative estimate of the potential of VOCs to induce food-finding behavior.

The ecological relevance of VOCs

Volatile organic compounds from cyanobacteria are known to be utilized as infochemicals involved in habitat-finding of aquatic insects (Evans 1982) and nematodes (Höckelmann et al. 2004). Furthermore, they possibly play a role in the oviposition-site finding of mosquitoes (Rejmankova et al. 2000). Interestingly, such ecological functions for VOCs in interspecific communication have so far been exclusively investigated for cyanobacteria. However, not only cyanobacteria, but also various eucaryotic algae are known to produce various odor compounds. For example, diatoms are known to produce straight chain and cyclic hydrocarbons, aldehydes and alcohols (Pohnert and Boland 1996, Wendel and Jüttner 1996), some of which are known to function as pheromones (i.e. for intraspecific communication) in brown algae (Müller et al. 1971). This can probably be explained by the phylogenetic relation between brown algae and diatoms (Pohnert and Boland 1996). However, to our knowledge, so far no results have been published either on the possible role of eucaryote-produced VOCs in interspecific communication, or on the VOCs released by (benthic) green algae. Especially the green algae seem to have been largely neglected in the context of biogenic volatiles, despite their considerable importance in the field (Stevenson et al. 1996).

This study gives the first indications that volatiles from green algae, released upon cell wounding, might serve as an important food-finding cue for freshwater benthic herbivores. Certainly, VOCs are not the only group of potential infochemicals liberated upon cell damage. Various other organic compounds such as sugars, amino acids and other low molecular weight carboxylic acids (such as propionic acid or butyric acid) have been described as attractants (Thomas et al. 1980, Thomas 1986) and feeding stimulants (Thomas et al. 1986, Thomas et al. 1989) in (tropical) freshwater snails. Some of these might also play a role for the chemical orientation of gastropod species from temperate latitudes. However, we did not find any attractance of butyric acid and (chironomid) carrion for *R. ovata* (P. Fink unpubl.). This can probably be explained by the fact that *R. ovata* feeds almost exclusively on periphyton and to a lesser degree on detritus (Calow 1970, Lodge 1986), but usually not on dead animals' tissue that could release significant amounts of dissolved carboxylic and amino acids. Furthermore, in the heterogeneous benthic environment, there could be many organisms and processes releasing such substances that do not necessarily indicate a food source to *R. ovata*. Therefore, algal volatiles, which are clearly distinguishable from diffuse sources of various compounds might be the more appropriate food-finding signal for *R. ovata*. Another point corroborating the effectiveness of VOCs as foraging kairomones is that in our experiments, the concentrations sufficient for a chemotactic response of the snails are about an order of magnitude lower than minimal concentrations effective for dissolved amino acids (Thomas et al. 1983). This is not surprising as, from terrestrial systems, it is known that the detection limit of invertebrates for volatile infochemicals is remarkably low (Harborne 1995). So far it remains unresolved, how the volatile infochemicals are released from benthic algae under natural conditions. However, in the field, a variety of mechanisms lead to lysis of benthic algal cells. Constantly, senescence and mechanical damage by hydrodynamic forces (Cattaneo 1990) occur. Furthermore, algal cells can become infected with parasitic fungi (Van Donk 1989) or viruses (Reisser 1993), resulting in increased cell lysis. Therefore, a constant "background" release of algal degradation products is to be expected in any natural biofilm community. Another potentially important release factor is the grazing activity of herbivores. Durrer and coworkers (Durrer et al. 1999) were able to show that grazing by herbivorous cladocera on planktonic cyanobacteria led to the release of significant amounts of VOCs both under field and laboratory conditions. Probably a similar mechanism could work in the benthos when grazing by one snail on an algal biofilm leads to rupture of algal cells and subsequently to an enhanced release of volatile lipoxygenase products, which would make it easier for other snails to detect this food patch. Hence, such behavioral response of gastropods to volatile lipoxygenase products (or oxylipins, Pohnert et al. 2002), could probably also help to explain the patchy occurrence of snails in lake littoral zones (Lodge 1986). These new findings on the role of green algal VOCs suggest that volatiles are not only important information transmitting cues in terrestrial ecosystems (Metcalf 1987), but also in (benthic) freshwater habitats.

Acknowledgements

We thank P. Merkel and E. Loher for excellent technical assistance with the VOC analyses and S. Boekhoff, B. Kumpfmüller and T. Basen for assistance with the food choice assays. W. Nagl gave valuable advice on the statistical analyses and L. Peters helped with the Figure on the experimental setup. We are indebted to M. Wolf for manufacturing the experimental containers. This study was supported by the Deutsche Forschungsgemeinschaft (DFG) within the Collaborative Research Centre SFB 454 – „Littoral of Lake Constance".

Chapter 7

Concluding remarks and perspectives

Littoral periphyton communities are among the most productive assemblages in aquatic ecosystems (Pinckney and Zingmark 1993). Therefore, they are also likely to constitute an important resource for benthic herbivores. An especially important group of herbivorous invertebrates are gastropod grazers, as they play a major role in structuring periphyton assemblages (e.g., Cattaneo 1983, Feminella and Hawkins 1995). In particular, gastropods have a rather high individual biomass in comparison to other invertebrate grazers (such as mayfly larvae) and thus, have a pronounced individual grazing impact, even at low densities. Nevertheless, herbivorous gastropods show a very high abundance in many lakes and rivers of temperate Europe. This is also the case in the prealpine central European Lake Constance, where gastropods have the second highest biomass among macroinvertebrates after the planktivorous bivalve *Dreissena polymorpha* (Baumgärtner 2004).
Furthermore, snails form an important link to higher trophic levels as they are an important food for various invertebrate (crayfish) and vertebrate (fish, waterfowl) predators (Turner et al. 2000). Thus, the group of herbivorous gastropods can be considered as a 'keystone link' in littoral ecosystems.

In this thesis, I investigated potential constraints for the biomass accrual of littoral gastropods. As an initial step, it was important to determine if stoichiometric constraints, that have been extensively studied in freshwater plankton could also play an important role for herbivorous benthic macroinvertebrates. Especially the consistent finding that planktonic herbivores maintain a relatively constant, i.e. homeostatic, body stoichiometry is a central point in the concept of ecological stoichiometry (Sterner and Elser 2002). However, so far it is largely unknown, whether this concept holds true for benthic herbivores in fresh waters. The data presented in **chapter 2** indicate that indeed, benthic herbivorous macroinvertebrates

maintain relatively constant body stoichiometries, similar to plankton organisms such as cladocera and copepods (Hessen and Lyche 1991). Another similarity to plankton ecosystems is the finding that pronounced taxon-specific differences in body stoichiometry are obvious. This suggests different requirements for nitrogen and phosphorous between different taxonomic groups of benthic herbivores. These differences could have a high impact on the competitive interactions between these species in an environment with changing nutrient availabilities. Of special interest is, if species that differ in their nutritional requirements for nitrogen and phosphorous (e.g. one species with lower nitrogen demand, but higher requirement for phosphorous and vice-versa for the other species) compete for periphyton varying in its relative content of both nutrients. Thus, competitive exclusion and coexistence scenarios could be tested as it has been done for freshwater phytoplankton (Tilman 1982) and zooplankton (Rothhaupt 1988).
Furthermore, the surprisingly high seasonal variation in periphyton carbon : nutrient ratios merits further attention. Particularly the high C:P ratios of periphyton in upper Lake Constance during the early summer months should be investigated in further detail (i.e. over several years), as it might indicate a period of strongly constrained food quality of the periphyton for littoral invertebrates. This corroborates recent findings that the concept of ecological stoichiometry (Sterner and Elser 2002) might also be applicable to littoral food webs (Frost et al. 2002b, Stelzer and Lamberti 2002, Hillebrand et al. 2004).

However, a mismatch between the elemental composition of a homeostatic herbivore and its resource does not imply a food quality limitation *per se*. On the contrary, such hypothetical constraints have to be subject to rigorous experimental testing. In particular, the extent to which dietary stoichiometry determines the food quality for benthic consumers should be determined. Furthermore, the proposed interaction between the amount of an available food or energy source (food quantity) and its nutritional value (food quality) has to be tested experimentally. These experimental studies should be performed under controlled laboratory conditions to exclude confounding effects that cannot be avoided under field conditions. This was done in the experiment presented in **chapter 3** with the snail *Radix ovata* fed different quantities of nutrient-saturated and nutrient-depleted algae. As predicted by theoretical models (Sterner 1997, Frost and Elser 2002a), there was a strong interaction between food quantity and quality. Interestingly, the effects of nutrient status of the algae (food quality) seemed to be even more important for the growth of juvenile *R. ovata* than the amount of available food (food quantity). Limiting amounts of high quality food supported higher shell growth rates of the snails than saturating quantities of nutrient-depleted algae. This highlights that food quality is a very important aspect that has to be considered in any quantitative evaluation of (benthic) food webs.

Another theoretical hypothesis in the context of ecological stoichiometry theory is the 'Growth Rate Hypothesis' (GRH). It predicts a negative correlation between an organism's growth rate and its body C:P ratio, as a high growth rate physiologically requires a high level of RNA, which is one of the biomolecules with the highest P-content (Elser et al. 2000a). There is an increasing amount of literature that the GRH is valid in terrestrial ecosystems as well as in freshwater plankton (Elser et al. 2003). However, again data from freshwater benthos is lacking. When the growth rates from the *R. ovata* growth experiment were plotted versus the animals' body C:P ratios, the observed pattern fits remarkably well to the negative relationship predicted by the GRH. Thus, the validity of this theoretical concept can probably be extended also to benthic ecosystems. This could be one of the mechanisms explaining the interacting effects of food quantity and food quality.

From an evolutionary perspective, it should be highly adaptive for herbivores to have compensation mechanisms that help them to cope with the highly variable nutrient composition of their resource. One such mechanism could be active selection of food with a high nutritional value, which is discussed below. Another mechanism could be compensatory feeding, i.e. an increased consumption of food items with low contents of essential nutrients in order to obtain sufficient amounts of that nutrient. Such a behavioural adaptation has been demonstrated for various groups of organisms and ecosystems (Raubenheimer 1992, Cruz-Rivera and Hay 2000). The data presented here suggest that *R. ovata* is also capable to increase the relative food consumption per grazer biomass to compensate for a low nutritional value of its diet. In the growth experiment, this compensatory feeding could slightly dampen, but not fully compensate the detrimental effects of the food's low nutrient content. However, compensatory feeding might have severe feedback effects on the whole littoral food web, when increased feeding of gastropods on periphyton low in P and/or N leads to a depletion of periphyton biomass. This could ultimately shift the benthic community from a P or N limitation to a C limitation. This can probably be one of the factors explaining the very low periphyton biomass during the summer months in the Lake Constance littoral. However, other factors such as physical disturbance and direct nutrient limitation of the periphyton community certainly play a role as well. This should be investigated in more detail as it will provide essential insights into the governing forces in freshwater periphyton communities.

Recently, research in food quality of algae for freshwater plankton focussed on a food quality constraint other than mineral nutrients. For the planktonic herbivore *Daphnia*, it seems that when molar C:P ratios in the seston are below a value of 300 (Sterner and Schulz 1998), the availability of polyunsaturated fatty acids (PUFAs) becomes the major constraint for *Daphnia* growth (Müller-Navarra 1995, Wacker and von Elert 2001). Similarly, the availability of PUFAs has a strong effect on different ontogenetic stages in the life cycle of the zebra mussel *Dreissena polymorpha* (Wacker et al. 2002, Wacker and von Elert 2002, Wacker and von Elert 2003, Wacker and von Elert 2004). Especially the group of n-3 PUFAs seems to be important, as this group cannot be synthesized by invertebrates (Stanley-Samuelson et al. 1988). In **chapter 4**, I investigated, whether the availability of the n-3 PUFAs α-linolenic acid (α-LA, C18:3 n-3) and eicosapentaenoic acid (EPA, C20:5 n-3) determines the food quality of two different algal species and a cyanobacterium for the growth of the freshwater gastropod *Bithynia tentaculata.* The method developed by (Von Elert 2002) was used to modify the fatty acid patterns of three species of primary producers. However, neither the absence of EPA (in the green alga *Scenedesmus obliquus*) nor a relatively low content of α-LA (in the diatom *Cyclotella meneghiniana*) constrained food quality of these algae for *B. tentaculata*. Hence, in contrast to the results obtained with *Daphnia* (Wacker and von Elert 2001) or the bivalve *Dreissena* (Wacker and von Elert 2002), no experimental evidence could be provided that PUFAs in general, and α-LA and EPA in particular, are limiting resources for the growth of this freshwater gastropod.

In contrast to *B. tentaculata*, which can switch between filter-feeding on phytoplankton and scraping of food organisms attached to the substrate with the radula, the pulmonate gastropod *Radix ovata* feeds exclusively on benthic food sources. Under field conditions, these are periphyton communities attached to hard substrates in the littoral. These benthic algal communities show strong variability in taxonomic (Swamikannu and Hoagland 1989, Stevenson et al. 1996) and nutrient composition (Kahlert 1998, Fink et al. submitted). As shown before, growth of *R. ovata* can be constrained by the availability of mineral nutrients. However, like for zooplankton herbivores (Sterner and Schulz 1998), when mineral nutrients are available in sufficient amounts, other cell constituents such as polyunsaturated fatty acids (PUFAs) might become limiting for this gastropod herbivore. Therefore, in **chapter 5**, I addressed the question, whether the PUFA composition of benthic algae commonly occurring in lake periphyton communities determines the growth of *R. ovata*. Additionally, the PUFA composition of the food algae might not only determine the algae's nutritive value for *R. ovata*, but also affect the PUFA composition of the snail's soft body.

The results of the *R. ovata* growth experiment are remarkably similar to those obtained with the prosobranch *B. tentaculata*. The different species of diatoms and green algae were of significantly different nutritional value for the growth of *R. ovata*, indicating that pronounced food quality differences do exist. However, neither the addition of α-LA to a diatom that is low in C18 fatty acids nor the supplementation of EPA to an EPA-free green alga had significant effects on the growth rates of juvenile *R. ovata*. Apparently these PUFAs are not limiting for this gastropod. However, the ability of *R. ovata* for biochemical conversion of these dietary PUFAs seems to be rather high, as even snails fed an α-LA-free or EPA-free diet still contained small amounts of α-LA or EPA, respectively, in their soft body tissues. *Daphnia* are known to be able to convert dietary α-LA into EPA by enzymatic elongation and desaturation (Von Elert 2002). Probably, *R. ovata* is not only able to convert α-LA into EPA, but also the other way round. This could allow the snails to partially compensate the absence of particular long chain PUFAs and be another important mechanism for these herbivores to cope with fluctuating contents of their periphyton resource. This ability for PUFA conversion in *R. ovata* can probably explain, why varying PUFA availabilities did not result in significant effects on the growth of juvenile *R. ovata*.

To summarize the results of the experiments on the role of PUFAs for the food quality of algae and cyanobacteria for freshwater gastropods: It becomes clear that the food's content of EPA is of minor importance for the growth of littoral snails. However, it was not yet tested if the reproduction of these gastropod molluscs could be influenced by the EPA content of the food as it has been demonstrated for the freshwater bivalve *Dreissena polymorpha* (Wacker and von Elert 2003, Wacker and von Elert 2004).

Furthermore, for both gastropod species investigated, an effect of α-LA on the growth of juvenile snails was indicated by the data, but in both cases not statistically significant. This might be due to the high variation between snails in the experimental treatments, resulting partially from the unknown previous conditions of the field-collected animals. However, in contrast to growth experiments with juvenile *D. polymorpha* (Wacker and von Elert 2002), where variance between (n = 3) replicates was extremely low, it was not possible to reduce the variance in the snail growth experiments despite a higher number (n = 5 for *R. ovata* and n = 8 for *B. tentaculata*) of replicates. The main reason is probably that genetically determined differences even between siblings originating from the same clutch are very high in freshwater snails (pers. observation) and hence, a reduction of the variance innate to the system is particularly difficult with these organisms. It cannot be excluded that further experiments would reveal a limitation of snail growth by the availability of α-LA and thus, the importance of PUFAs for the performance of freshwater gastropods is not yet fully clear.

As mentioned above, one way for gastropods to cope with highly fluctuating nutrient availabilities might be the active selection of patches with high quality food (Butler et al. 1989). For organisms such as snails which have rather low visual capabilities and for which locomotion is associated with high energetic costs, it should therefore be adaptive to make use of foraging kairomones to identify sources of food already from a distance and then actively move towards this source (chemotaxis). Such foraging kairomones (sensu Ruther et al. 2002) are very commonly found in terrestrial plant-herbivore interactions, especially for insects (Metcalf 1987), but also for snails (Croll and Chase 1980). However, reports on chemically mediated food finding over distance are scarce for aquatic ecosystems (but see (Van Gool and Ringelberg 1996)).

The group of volatile organic compounds (VOCs) may serve as important infochemicals, especially in biofilms of benthic algae and cyanobacteria (Watson 2003). Benthic mat-forming cyanobacteria have been shown to produce volatiles that are used as habitat-finding cues by insects and nematodes (Evans 1982,

Höckelmann et al. 2004). Thus, we hypothesized that algal-derived VOCs might also serve as foraging kairomones for herbivorous freshwater gastropods. This was tested using a model system consisting of the gastropod periphyton grazer *Radix ovata* and the mat-forming benthic green alga *Ulothrix fimbriata*.

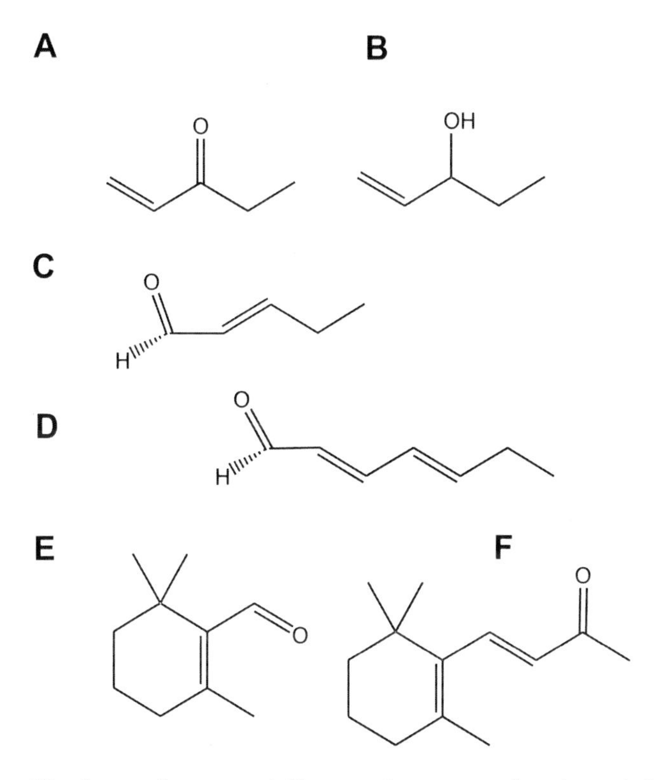

Figure 1: Structures of some volatile organic compounds released from *U. fimbriata* upon cell damage; A) 1-Penten-3-one; B) 1-Penten-3-ol; C) 2-Pentenal; D) 2,4-Heptadienal; E) β-Cyclocitral F) β-Ionone.

To test for the proposed attractivity of algae-derived volatiles to *R. ovata*, I developed a new choice assay to determine the behavioural response of juvenile snails to algal odours. In particular, it allowed for separating the effects of food-derived odours and the food source itself to disentangle effects of long-range (odour) and short-range (taste) chemoreception. As outlined in **chapter 6**, *U. fimbriata* released a diverse bouquet of VOCs upon cell damage, some of which are depicted in Fig. 1 to exemplify the structural diversity of liberated compounds. When these volatiles were trapped on an adsorbent and the eluate of the adsorbent offered to juvenile *R. ovata* in the choice assay, the extract of VOCs released from damaged *U. fimbriata* cells was clearly preferred over a control extract. Hence, I could unambiguously demonstrate that VOCs are responsible for this attractivity.

Further assays revealed that a multicomponent odour, rather than a single compound, is responsible for this attractivity and the activity could be assigned to a mixture of four major compounds (1-penten-3-one, 1-penten-3-ol, 2-pentenal and 2,4-heptadienal). So far, biogenic volatiles have primarily been seen as important information transmitting cues in terrestrial ecosystems (Metcalf 1987, Ruther et al. 2002). This study shows, that they might also function as foraging kairomones in freshwater benthic habitats.

During the last century, research on the ecology of lentic ecosystems has been largely biased towards the pelagic components. A lot of progress in ecological theory in general was developed in pelagic systems which are comparably easy to manipulate in the laboratory and to model mathematically. This is rather more difficult for the physically highly structured benthic habitats.

In this thesis, I investigated some of the key issues of recent research in freshwater plankton for their relevance to littoral food webs. By using the keystone group of gastropod herbivores, it was possible to get insights into the differences and similarities between the physically quite different, but inevitably interconnected pelagic and littoral habitats in lake ecosystems.

Summary

Freshwater gastropods form a key link in the food webs of lake littoral zones. They make benthic primary production available to invertebrate and vertebrate predators. However, their community biomass is probably not only influenced by top-down effects from predators, but also by bottom-up effects from their periphyton resource. Especially the nutritional quality of periphytic algae might constrain the growth of juvenile snails. However, it is unclear, which factors determine the nutritional quality of benthic primary producers for freshwater gastropods. Among these factors could be compounds within the cells of the primary producers that are essential for the herbivores. In the present thesis, I investigated several of these factors that potentially determine the nutritional quality of periphyton for gastropod grazers. Surveys were performed both in the field and under standardised conditions in laboratory growth experiments.

Nitrogen (N) and phosphorus (P) are considered to be keystone nutrients that control secondary production in freshwater ecosystems; insufficient availability of N and P are known to limit herbivore growth in the pelagic zone of oligotrophic lakes. Planktonic herbivores maintain a homeostatic, i. e. relatively constant body C:N:P ratio. I showed here, that also benthic herbivores maintain relatively homeostatic body C:N:P ratios despite considerable temporal and spatial variation in the C:N:P ratios of their periphyton resource. Furthermore, to what extent the dietary stoichiometry determines the food quality for benthic consumers is not clear. In this thesis, I investigated the potential of primary producers low in P, N, or P and N to constrain growth of the freshwater gastropod *Radix ovata*. The results indicated that a low availability of N and especially P can strongly limit the growth of juvenile snails. The snails responded to the low nutritional quality in their food source by an increased nutrient retention during the digestive process and by a compensatory increase in feeding rates. This could partially dampen, but not remove the detrimental effect of low N and P availability on the snails' growth.

Besides the mineral nutrients like N and P, also biochemical constituents of primary producers can be essential for herbivores if they are not able to synthesize these compounds in sufficient amounts to meet their physiological needs. Especially polyunsaturated fatty acids (PUFAs) seem to be important for the efficient transfer of photosynthetically fixed carbon to higher trophic levels, as it was previously shown for filter-feeding cladocera and zebra mussels. In how far the availability of dietary PUFAs is also crucial for the growth of freshwater snails, was investigated in two experimental studies with gastropod species differing in their feeding behavior. The snails were fed algae and cyanobacteria with experimentally modified fatty acid patterns to investigate the role of single PUFAs in determining the food quality of primary producers for the gastropods. Furthermore, the fatty acid content of algae, snail soft bodies and the snails' faecal pellets was determined quantitatively using gas chromatography. Data presented here suggest, that the availability of the n-3 PUFAs α-linolenic acid and eicosapentaenoic acid was not among the main factors responsible for the observed differences in food quality. Hence, results presented in this thesis suggest that for the nutritional quality of periphyton for snails, the algae's content of mineral nutrients as nitrogen and especially phosphorous are more important than the amount of polyunsaturated fatty acids supplied to the grazers by the periphyton resource.

The distribution of periphyton in littoral habitats is highly patchy with respect to biomass, taxonomy and nutrient status of the algae and cyanobacteria contributing to this community of benthic primary producers. Therefore, it should be highly adaptive for gastropod herbivores to utilise chemical cues in order to detect patches of (high quality) food. To investigate this, I developed a new behavioural assay system to test for the food-finding behaviour of aquatic snails. In particular, the assay was designed to allow for a separation of odour mediated food-finding (working on longer distances) from taste-mediated food preference (working only by direct contact with the food). Furthermore, by applying gas chromatography / mass spectrometry trace analysis, I was able to quantitatively determine the bouquet of volatile organic compounds released upon cell damage from the filamentous green alga *Ulothrix fimbriata*. By offering both *U. fimbriata* biomass and extracts of *U. fimbriata* volatiles in the behavioural assay system, I could show that the freshwater gastropod *R. ovata* is capable of a behavioural, chemotactic response to food-derived odours. This infochemical-mediated foraging strategy allows the snails to save valuable energy and resources in finding patchily distributed food in their highly heterogeneous littoral habitat. In particular, volatile organic compounds, released from benthic algae, can be used by the snails as foraging kairomones that indicate the presence of a food patch. These adaptations probably contribute to the extreme success of freshwater gastropods in many lake and river littoral zones.

Zusammenfassung

Süßwasserschnecken haben eine Schlüsselstellung im Nahrungsnetz von Seeuferzonen inne. Sie machen die benthische Primärproduktion für invertebrate und vertebrate Prädatoren verfügbar. Jedoch ist die Biomasse der Schneckengemeinschaft nicht nur von „top-down" Effekten durch Räuber beeinflusst, sondern auch durch „bottom-up" Effekte durch ihre Ressource, das Periphyton. Insbesondere die Nahrungsqualität periphytischer Algen könnte das Wachstum junger Schnecken begrenzen. Es ist dabei aber unklar, welche Faktoren die Nahrungsqualität benthischer Primärproduzenten für Süßwasserschnecken bestimmen. Unter diesen Faktoren könnten auch Zellinhaltsstoffe der Primärproduzenten sein, die für die Herbivoren essentiell sind. In dieser Arbeit habe ich mehrere dieser Faktoren, die möglicherweise die Nahrungsqualität von Periphyton für Schnecken-Weidegänger bestimmen, untersucht. Die Untersuchungen wurden sowohl im Freiland, als auch unter standardisierten Bedingungen in Labor-Wachstumsversuchen durchgeführt.

Stickstoff (N) und Phosphor (P) werden als Schlüsselnährstoffe gesehen, die die Menge der Sekundärproduktion in Süßwasserökosystemen bestimmen; eine zu geringe Verfügbarkeit von N und P kann bekannterweise das Wachstum von Herbivoren im Pelagial oligotropher Seen limitieren. Planktische Herbivore erhalten ein homöostatisches, d. h. relativ konstantes C:N:P Verhältnis in ihren Körpern. Hier konnte ich zeigen, dass benthische Herbivore trotz beachtlicher zeitlicher und räumlicher Variationen im C:N:P Verhältnis ihrer Ressource, des Periphytons, ein relativ konstantes C:N:P Verhältnis in ihren Körpern erhalten. Des Weiteren ist nicht klar, inwieweit die Stöchiometrie des Futters dessen Futterqualität für benthische Konsumenten bestimmt. In dieser Arbeit untersuchte ich das Potential von P-verarmten, N-verarmten oder P- und N-verarmten Primärproduzenten das Wachstum der Süßwasserschnecke *Radix ovata* zu beschränken. Die Ergebnisse deuten darauf hin, dass eine geringe Verfügbarkeit von N und besonders P das Wachstum juveniler Schnecken stark limitieren kann. Als Reaktion auf die geringe Nahrungsqualität ihrer Futterquelle erhöhten die Schnecken ihre Nährstoffaufnahme während des Verdauungsprozesses und antworteten durch eine kompensatorische Erhöhung der Freßraten. Dies konnte die abträglichen Effekte der geringen N- und P-Verfügbarkeit auf das Wachstum der Schnecken zwar dämpfen, aber nicht entfernen.

Neben mineralischen Nährstoffen wie N und P können auch biochemische Bestandteile von Primärproduzenten essentiell für Herbivore sein, wenn diese nicht in der Lage sind, solche Bestandteile in genügender Menge selbst herzustellen um ihren physiologischen Anforderungen zu genügen. Insbesondere mehrfach ungesättigte Fettsäuren (PUFAs) scheinen wichtig für den effizienten Transfer von potosynthetisch fixiertem Kohlenstoff zu höheren trophischen Ebenen zu sein, wie für filtrierende Cladoceren und Dreikantmuscheln gezeigt werden konnte. Inwieweit die Verfügbarkeit von PUFAs in der Nahrung auch für das Wachstum von Süßwasserschnecken entscheidend ist, wurde in zwei experimentellen Studien mit Schnecken mit unterschiedlichen Ernährungsweisen untersucht. Die Schnecken wurden mit Algen und Cyanobakterien mit experimentell modifizierten Fettsäuremustern gefüttert, um die Rolle einzelner PUFAs für die Nahrungsqualität der Primärproduzenten für die Schnecken zu untersuchen. Des Weiteren wurden die Fettsäuregehalte der Algen, der Schneckenweichkörper und der Fäzespartikel der Schnecken quantitativ mittels Gaschromatographie bestimmt. Die hier gezeigten Daten deuten darauf hin, dass die Verfügbarkeit der n-3 PUFAs α-Linolensäure und Eicosapentaensäure nicht zu den Hauptfaktoren gehört, die für die beobachteten Unterschiede in der Nahrungsqualität verantwortlich sind. Die Ergebnisse dieser Arbeit deuten also darauf hin, dass für die Nahrungsqualität von Periphyton für Schnecken, der Gehalt der Algen an mineralischen Nährstoffen wie Stickstoff und besonders Phosphor wichtiger ist, als die Menge an mehrfach ungesättigten Fettsäuren, die dem Weidegänger durch das Periphyton zugeführt wird.

Die Verteilung des Periphytons in Uferhabitaten ist sehr ungleichmäßig, sowohl im Bezug auf Biomasse, taxonomische Zusammensetzung und Nährstoffstatus der Algen und Cyanobakterien, die zu dieser Gemeinschaft von Primärproduzenten beitragen. Deshalb sollte es hoch adaptiv für herbivore Schnecken sein, chemische Signalstoffe zu nutzen, um Stellen mit (qualitativ hochwertigem) Futter zu finden. Um dies zu untersuchen, habe ich ein neues Verhaltensbiotest-System entwickelt, um das Futterfindungsverhalten aquatischer Schnecken zu testen. Insbesondere wurde der Biotest darauf hin entwickelt, eine Unterscheidung von durch Geruchsstoffe vermittelter Futtersuche (auf längere Distanzen) von geschmacksvermittelter Futterpräferenz (die einen direkten Kontakt mit dem Futter erfordert) zu trennen. Des Weiteren war ich durch die Anwendung von gaschromatographisch-massenspektrometrischer Spurenanalytik in der Lage, das Bukett flüchtiger organischer Substanzen, das bei Beschädigung der Zellen aus der filamentösen Grünalge *Ulothrix fimbriata* freigesetzt wird, quantitativ zu bestimmen. Indem ich im Verhaltensbiotest sowohl Biomasse von *U. fimbriata*, als auch Extrakte von flüchtigen Substanzen aus *U. fimbriata* anbot, konnte ich zeigen, dass die Süßwasserschnecke *R. ovata* zu einer chemotaktisch vermittelten Verhaltensantwort auf futterbürtige Signalstoffe in der Lage ist. Diese durch Infochemikalien vermittelte Strategie der Nahrungssuche erlaubt es den Schnecken, wertvolle Energie und Ressourcen beim Auffinden ungleichmäßig verteilter Nahrung in ihrem hochgradig heterogenen Uferhabitat zu sparen. Insbesondere können flüchtige organische Verbindungen, die von benthischen Algen freigesetzt werden, von den Schnecken als Fouragierkairomone genutzt werden, die das Vorhandensein von Futter anzeigen. Diese Anpassungen tragen wahrscheinlich zum extrem großen Erfolg von Süßwasserschnecken in vielen See- und Flußuferbereichen bei.

Literature cited

Acharya, K., M. Kyle and J. J. Elser. 2004. Biological stoichiometry of *Daphnia* growth: An ecophysiological test of the growth rate hypothesis. Limnol. Oceanogr. **49:** 656-665.

Andersen, T. and D. O. Hessen. 1991. Carbon, nitrogen, and phosphorus content of freshwater zooplankton. Limnol. Oceanogr. **36:** 807-814.

Anderson, M. J. and A. J. Underwood. 1997. Effects of gastropod grazers on recruitment and succession of an estuarine assemblage: A multivariate and univariate approach. Oecologia **109:** 442-453.

Barnese, L. E., R. L. Lowe and R. D. Hunter. 1990. Comparative Grazing Efficiency of Pulmonate and Prosobranch Snails. J. N. Am. Benthol. Soc. **9:** 35-44.

Baumgärtner, D. 2004. Principles of macroinvertebrate community structure in the littoral zone of Lake Constance. PhD Thesis. University of Konstanz, Germany.

Brendelberger, H. 1995a. Growth of juvenile *Bithynia tentaculata* (Prosobranchia, Bithyniidae) under different food regimes: A long-term laboratory study. J. Moll. Stud. **61:** 89-95.

Brendelberger, H. 1995b. Dietary preference of three freshwater gastropods for eight natural foods of different energetic content. Malacologia **36:** 147-153.

Brendelberger, H. 1997a. Contrasting feeding strategies of two freshwater gastropods, *Radix peregra* (Lymnaeidae) and *Bithynia tentaculata* (Bithynidae). Arch. Hydrobiol. **140:** 1-21.

Brendelberger, H. 1997b. Determination of digestive enzyme kinetics: A new method to define trophic niches in freshwater snails. Oecologia **109:** 34-40.

Brendelberger, H. and S. Jürgens. 1993. Suspension feeding in *Bithynia tentaculata* (Prosobranchia, Bithyniidae), as affected by body size, food and temperature. Oecologia **94:** 36-42.

Brett, M. T. and D. C. Müller-Navarra. 1997. The role of highlyunsaturated fatty acids in aquatic foodweb processes. Freshw. Biol. **38:** 438-499.

Butler, N. M., C. A. Suttle and W. E. Neill. 1989. Discrimination by fresh-water zooplankton between single algal cells differing in nutritional status. Oecologia **78:** 368-372.

Calow, P. 1970. Studies on the natural diet of *Lymnaea pereger obtusa* (Kobelt) and its possible ecological implications. Proc. malac. Soc. Lond. **39:** 203-215.

Calow, P. and L. J. Calow. 1975. Cellulase activity and niche separation in freshwater gastropods. Nature **255:** 478-480.

Carrillo, P., I. Reche and L. Cruz-Pizarro. 1996. Intraspecific stoichiometric variability and the ratio of nitrogen to phosphorus resupplied by zooplankton. Freshw. Biol. **36:** 363-374.

Cattaneo, A. 1983. Grazing on epiphytes. Limnol. Oceanogr. **28:** 124-132.

Cattaneo, A. 1990. The effect of fetch on periphyton spatial variation. Hydrobiologia **206:** 1-10.

Cattaneo, A. and J. Kalff. 1986. The effect of grazer size manipulation on periphyton communities. Oecologia **69:** 612-617.

Clarke, K. R. and R. M. Warwick. 2001. Changes in marine communities: an approach to statistical analysis and interpretation, PRIMER-E, Plymouth, U.K.

Cobelas, M. A. and J. Z. Lechardo. 1988. Lipids in microalgae: a review I. Biochemistry. Grasas y Aceites **40:** 118-145.

Croll, R. P. 1983. Gastropod chemoreception. Biol. Rev. **58:** 293-319.

Croll, R. P. and R. Chase. 1980. Plasticity of olfactory orientation to foods in the snail *Achatina fulica*. J. Comp. Physiol. **136:** 267-277.

Cruz-Rivera, E. and M. E. Hay. 2000. Can quantity replace quality? Food choice, compensatory feeding, and fitness of marine mesograzers. Ecology **81:** 201-219.

Darchambeau, F., P. J. Faerovig and D. O. Hessen. 2003. How *Daphnia* copes with excess carbon in its food. Oecologia **136:** 336-346.

Delaunay, F., Y. Marty, J. Moal and J. F. Samain. 1993. The effect of monospecific algal diets on growth and fatty acid Composition of *Pecten maximus* (L) larvae. J. Exp. Mar. Biol. Ecol. **173:** 163-179.

DeMott, W. R. 2003. Implications of element deficits for zooplankton growth. Hydrobiologia 177-184.

DeMott, W. R., R. D. Gulati and K. Siewertsen. 1998. Effects of phosphorus-deficient diets on the carbon and phosphorus balance of *Daphnia magna*. Limnol. Oceanogr. **43:** 1147-1161.

DeMott, W. R., R. D. Gulati and E. Van Donk. 2001. *Daphnia* food limitation in three hypereutrophic Dutch lakes: Evidence for exclusion of large-bodied species by interfering filaments of cyanobacteria. Limnol. Oceanogr. **46:** 2054-2060.

Denny, M. 1980. Locomotion: the cost of Gastropod crawling. Science **208:** 1288-1290.

Diaz Villanueva, V. D., R. Albarino and B. Modenutti. 2004. Grazing impact of two aquatic invertebrates on periphyton from an Andean-Patagonian stream. Arch. Hydrobiol. **159:** 455-471.

Droop, M. R. 1974. The nutrient status of algal cells in continuous culture. J. mar. biol. Ass. U.K. **54:** 825-855.

Durrer, M., U. Zimmermann and F. Jüttner. 1999. Dissolved and particle-bound geosmin in a mesotrophic lake (Lake Zurich): Spatial and seasonal distribution and the effect of grazers. Wat. Res. **33:** 3628-3636.

El-Sayed, A., M. Bengtsson, S. Rauscher, J. Lofqvist and P. Witzgall. 1999. Multicomponent sex pheromone in codling moth (Lepidoptera : Tortricidae). Env. Entomol. **28:** 775-779.

Elser, J. J., E. R. Marzolf and C. R. Goldman. 1990. Phosphorus and nitrogen limitation of phytoplankton growth in the freshwaters of North America: a review and critique of experimental enrichments. Can. J. Fish. Aquat. Sci. **47:** 1468-1477.

Elser, J. J., D. R. Dobberfuhl, N. A. MacKay and J. H. Schampel. 1996. Organism size, life history and N:P stoichiometry - Toward a unified view of cellular and ecosystem processes. BioScience **46:** 674-684.

Elser, J. J., R. W. Sterner, E. Gorokhova, W. F. Fagan, T. A. Markow, J. B. Cotner, J. F. Harrison, S. E. Hobbie, G. M. Odell and L. J. Weider. 2000a. Biological stoichiometry from genes to ecosystems. Ecol. Lett. **3:** 540-550.

Elser, J. J., W. F. Fagan, R. F. Denno, D. R. Dobberfuhl, A. Folarin, A. Huberty, S. Interlandi, S. S. Kilham, E. Mc Cauley, K. L. Schulz, E. H. Siemann and R. W. Sterner. 2000b. Nutritional constraints in terrestrial and freshwater food webs. Nature **408:** 578-580.

Elser, J. J., K. Acharya, M. Kyle, J. Cotner, W. Makino, T. Markow, T. Watts, S. Hobbie, W. Fagan, J. Schade, J. Hood and R. W. Sterner. 2003. Growth-rate stoichiometry couplings in diverse biota. Ecol. Lett. **6:** 936-943.

Evans, W. G. 1982. *Oscillatoria* sp. (Cyanophyta) mat metabolites implicated in habitat selection in *Bembidion obtusidens* (Coleoptera: Carabidae). J. Chem. Ecol. **8:** 671-678.

Færøvig, P. J. and D. O. Hessen. 2003. Allocation strategies in crustacean stoichiometry: the potential role of phosphorus in the limitation of reproduction. Freshw. Biol. **48:** 1782-1792.

Feminella, J. W. and C. Hawkins, P. 1995. Interactions between stream herbivores and periphyton: a quantitative analysis of past experiments. J. N. Am. Benthol. Soc. **14:** 465-509.

Fink, P. and E. Von Elert. *in press*. Food quality of algae and cyanobacteria for the freshwater gastropod *Bithynia tentaculata*: the role of polyunsaturated fatty acids. Verh. Int. Verein. Limnol. **29:**

Fink, P. and E. Von Elert. *submitted*. Stoichiometric constraints in benthic food webs: nutrient limitation and compensatory feeding in the freshwater snail *Radix ovata. Submitted to Oikos.*

Fink, P., L. Peters and E. Von Elert. *in press*. Stoichiometric mismatch between littoral invertebrates and their periphyton food. Arch. Hydrobiol.

Fitzgerald, G. P. and T. C. Nelson. 1966. Extractive and enzymatic analyses for limiting or surplus phosphorus in algae. J. Phycol. **2:** 32-37.

Frost, P. C. and J. J. Elser. 2002a. Growth responses of littoral mayflies to the phosphorus content of their food. Ecol. Lett. **5:** 232-240.

Frost, P. C. and J. J. Elser. 2002b. Effects of light and nutrients on the net accumulation and elemental composition of epilithon in boreal lakes. Freshw. Biol. **47:** 173-183.

Frost, P. C., J. J. Elser and M. A. Turner. 2002a. Effects of caddisfly grazers on the elemental composition of epilithon in a boreal lake. J. N. Am. Benthol. Soc. **21:** 54-63.

Frost, P. C., R. S. Stelzer, G. A. Lamberti and J. J. Elser. 2002b. Ecological stoichiometry of trophic interactions in the benthos: Understanding the role of C:N:P ratios in lentic and lotic habitats. J. N. Am. Benthol. Soc. **21:** 515-528.

Frost, P. C., S. E. Tank, M. A. Turner and J. J. Elser. 2003. Elemental composition of littoral invertebrates from oligotrophic and eutrophic Canadian lakes. J. N. Am. Benthol. Soc. **22:** 51-62.

Gal, J., M. V. Bobkova, V. V. Zhukov, I. P. Shepeleva and V. B. Meyer-Ruchow. 2004. Fixed focal-length optics in pulmonate snails (Mollusca, Gastropoda): squaring phylogenetic background and ecophysiological needs (II). Inv. Biol. **123:** 116-127.

Greenberg, A. E., R. R. Trussel and L. S. Clesceri. 1985. Standard methods for the examination of water and wastewater. American Public Health Association (APHA), Washington D.C., USA.

Guillard, R. R. L. 1975. Culture of phytoplankton for feeding marine invertebrates. p. 29-60. *In* W. L. Smith, and Chanley, M.H. [eds.], Culture of marine invertebrate animals. Plenum Press.

Guillard, R. R. L. and C. J. Lorenzen. 1972. Yellow-green algae with chlorophyllide c. J. Phycol. **8:** 10-14.

Halitschke, R., J. Ziegler, M. Keinänen and I. T. Baldwin. 2004. Silencing of hydroperoxide lyase and allene oxide synthase reveals substrate and defense signalling crosstalk in *Nicotiana attenuata*. Plant J. **40:** 35-46.

Harborne, J. B. 1995. Ökologische Biochemie. Eine Einführung, Spektrum Akademischer Verlag.

Hessen, D. O. 1990. Carbon, nitrogen and phosphorus status in *Daphnia* at varying food conditions. J. Plankton Res. **12:** 1239-1249.

Hessen, D. O. and A. Lyche. 1991. Inter- and intraspecific variations in zooplankton element composition. Arch. Hydrobiol. **121:** 343-353.

Hillebrand, H. and U. Sommer. 1999. The nutrient stoichiometry of benthic microalgal growth: Redfield proportions are optimal. Limnol. Oceanogr. **44:** 440-446.

Hillebrand, H. and M. Kahlert. 2001. Effect of grazing and nutrient supply on periphyton biomass and nutrient stoichiometry in habitats of different productivity. Limnol. Oceanogr. **46:** 1881-1898.

Hillebrand, H., G. de Montpellier and A. Liess. 2004. Effects of macrograzers and light on periphyton stoichiometry. Oikos **106:** 93-104.

Höckelmann, C., T. Moens and F. Jüttner. 2004. Odor compounds from cyanobacterial biofilms acting as attractants and repellents for free-living nematodes. Limnol. Oceanogr. **49:** 1809-1819.

Ianora, A., A. Miralto, S. A. Poulet, Y. Carotenuto, I. Buttino, G. Romano, R. Casotti, G. Pohnert, T. Wichard, L. Colucci-D'Amato, G. Terrazzano and V. Smetacek.

2004. Aldehyde suppression of copepod recruitment in blooms of a ubiquitous planktonic diatom. Nature **429:** 403-407.

Jüttner, F. 1987. Volatile organic substances. p. 453-469. *In* P. Fay and C. Van Baalen [eds.], The Cyanobacteria. Elsevier.

Jüttner, F. 1988a. Biochemistry of biogenic off-flavour compounds in surface waters. Wat. Sci. Technol. **20:** 107-116.

Jüttner, F. 1988b. Quantitative trace analysis of volatile organic compounds. Meth. .Enzymol. **167:** 609-616.

Jüttner, F. 2001. Liberation of 5,8,11,14,17-eicosapentaenoic acid and other polyunsaturated fatty acids from lipids as a grazer defense reaction in epilithic diatom biofilms. J. Phycol. **37:** 744-755.

Jüttner, F. and U. Dürst. 1997. High lipoxygenase activities in epilithic biofilms of diatoms. Arch. Hydrobiol. **138:** 451-463.

Jüttner, F., J. Leonhardt and S. Möhren. 1983. Environmental factors affecting the formation of mesityloxide, dimethylallylic alcohol and other volatile compounds excreted by *Anabaena cylindrica*. J. Gen. Microbiol. **129:** 407-412.

Kahlert, M. 1998. C:N:P ratios of freshwater benthic algae. Arch. Hydrobiol. Spec. Issues Advanc. Limnol. **51:** 105-114.

Kahlert, M., A. T. Hasselrot, H. Hillebrand and K. Pettersson. 2002. Spatial and temporal variation in the biomass and nutrient status of epilithic algae in Lake Erken, Sweden. Freshw. Biol. **47:** 1191-1215.

Kessler, A. and I. T. Baldwin. 2001. Defensive function of herbivore-induced plant volatile emissions in nature. Science **291:** 2141-2144.

Lampert, W. 1981. Inhibitory and toxic effects of blue-green algae on *Daphnia*. Int. Revue ges. Hydrobiol. **66:** 285-298.

Lampert, W. and I. Trubetskova. 1996. Juvenile growth rate as a measure of fitness in *Daphnia*. Funct. Ecol. **10:** 631-635.

Liess, A. and H. Hillebrand. 2004. Invited review: Direct and indirect effects in herbivore periphyton interactions. Arch. Hydrobiol. **159:** 433-453.

Liess, A. and H. Hillebrand. 2005. Stoichiometric variation in C : N, C : P, and N : P ratios of littoral benthic invertebrates. J. N. Am. Benthol. Soc. **24:** 256.

Lodge, D. M. 1986. Selective grazing on periphyton: a determinant of freshwater gastropod microdistributions. Freshw. Biol. **16:** 831-841.

Madsen, H. 1992. Food Selection by Freshwater Snails in the Gezira Irrigation Canals Sudan. Hydrobiologia **228:** 203-217.

Metcalf, R. L. 1987. Plant volatiles as insect attractants. CRC Crit. Rev. Plant Sci. **5:** 251-301.

Miralto, A., G. Barone, G. Romano, S. A. Poulet, A. Ianora, G. L. Russo, I. Buttino, G. Mazzarella, M. Laabir, M. Cabrini and M. G. Glacobbe. 1999. The insidious effect of diatoms on copepod reproduction. Nature **402:** 173-176.

Moog, O. 1995. Fauna aquatica austriaca - Einstufungskatalog benthischer Evertebraten Österreichs. Österreichisches Bundesministerium für Land- und Forstwirtschaft. Wien, Austria.

Müller, D. G., L. Jaenicke, M. Donike and T. Akintobi. 1971. Sex attractant in a brown alga - chemical structure. Science **171:** 815.

Müller-Navarra, D. 1995. Evidence that a highly unsaturated fatty acid limits *Daphnia* growth in nature. Arch. Hydrobiol. **132:** 297-307.

Peters, L., N. Scheifhacken, M. Kahlert and K.-O. Rothhaupt. 2005. A precise in situ method for sampling periphyton communities in lakes and streams. Arch. Hydrobiol. **163:** 133-141.

Pillsbury, K. S. 1985. The relative food value and biochemical composition of 5 phytoplankton diets for Queen Conch, *Strombus gigas* (Linné) larvae. J. Exp. Mar. Biol. Ecol. **90:** 221-231.

Pinckney, J. L. and R. G. Zingmark. 1993. Modeling the annual production of intertidal benthic microalgae in estuarine ecosystems. J. Phycol. **29:** 396-407.

Plath, K. and M. Boersma. 2001. Mineral limitation of zooplankton: Stoichiometric constraints and optimal foraging. Ecology **82:** 1260-1269.

Pohnert, G. 2000. Wound-activated chemical defense in unicellular planktonic algae. Angew. Chem. Int. Ed. **39:** 4352-4354.

Pohnert, G. 2002. Phospholipase A2 activity triggers the wound-activated chemical defense in the diatom *Thalassiosira rotula*. Plant Physiol. **129:** 103-111.

Pohnert, G. and W. Boland. 1996. Biosynthesis of the algal pheromone hormosirene by the freshwater diatom *Gomphonema parvulum* (Bacillariophyceae). Tetrahedron **52:** 10073-10082.

Pohnert, G., O. Lumineau, A. Cueff, S. Adolph, C. Cordevant, M. Lange and S. Poulet. 2002. Are volatile unsaturated aldehydes from diatoms the main line of chemical defence against copepods? Mar. Ecol. Prog. Ser. **245:** 33-45.

Porter, K. G. 1975. Viable gut passage of gelatinous green algae ingested by *Daphnia*. Verh. Internat. Verein. Limnol. **19:** 2840-2850.

Raubenheimer, D. 1992. Tannic acid, protein, and digestible carbohydrate: dietary imbalance and nutritional compensation in locusts. Ecology **73:** 1012-1027.

Reisser, W. 1993. Viruses and virus-like particles of fresh-water and marine eukaryotic algae - a review. Arch. Protistenkd. **143:** 257-265.

Rejmankova, E., R. M. Higashi, D. R. Roberts, M. Lege and R. G. Andre. 2000. The use of solid phase micro-extraction (SPME) devices in analysis for potential mosquito oviposition attractant chemicals from cyanobacterial mats. Aquat. Ecol. **34:** 413-420.

Rosemond, A. D., P. J. Mulholland and S. H. Brawley. 2000. Seasonally shifting limitation of stream periphyton: Response of algal populations and assemblage biomass and productivity to variation in light, nutrients, and herbivores. Can. J. Fish. Aquat. Sci. **57:** 66-75.

Rothhaupt, K. O. 1988. Mechanistic resource competition theory applied to laboratory experiments with zooplankton. Nature **333:** 660-662.

Ruther, J., T. Meiners and J. L. M. Steidle. 2002. Rich in phenomena - lacking in terms. A classification of kairomones. Chemoecology **12:** 161-167.

Simkin, A. J., S. H. Schwartz, M. Auldridge, M. G. Taylor and H. J. Klee. 2004. The tomato carotenoid cleavage dioxygenase 1 genes contribute to the formation of the flavor volatiles beta-ionone, pseudoionone, and geranylacetone. Plant J. **40:** 882-892.

Speiser, B. and M. Rowell-Rahier. 1991. Effects of food availability nutritional value and alkaloids on food choice in the generalist herbivore *Arianta arbustorum* (Gastropoda, Helicidae). Oikos **62:** 306-318.

Stanley-Samuelson, D. W., R. A. Jurenka, C. Cripps, G. J. Blomquist and M. Derenobales. 1988. Fatty acids in insects - composition, metabolism, and biological significance. Arch. Insect Biochem. Physiol. **9:** 1-33.

StatSoft. 2004. STATISTICA (data analysis software system). version 6. www.statsoft.com.

Steinman, A. D. 1996. Effects of grazers on freshwater benthic algae. p. 341-373. *In* R. J. Stevenson, M. L. Bothwell and R. L. Lowe [eds.], Algal ecology - Freshwater benthic ecosystems. Academic Press.

Stelzer, R. S. and G. A. Lamberti. 2002. Ecological stoichiometry in running waters: periphyton chemical composition and snail growth. Ecology **83:** 1039-1051.

Sterner, R. W. 1990. The ratio of nitrogen to phosphorus resupplied by herbivores: zooplankton and the algal competitive arena. Am. Nat. **136:** 209-229.

Sterner, R. W. 1993. *Daphnia* growth on varying quality of *Scenedesmus*: mineral limitation of zooplankton. Ecology **74:** 2351-2360.

Sterner, R. W. 1997. Modelling interactions of food quality and quantity in homeostatic consumers. Freshw. Biol. **38:** 473-481.

Sterner, R. W. and D. O. Hessen. 1994. Algal nutrient limitation and the nutrition of aquatic herbivores. Ann. Rev. Ecol. Syst. **25:** 1-29.

Sterner, R. W. and K. L. Schulz. 1998. Zooplankton nutrition: Recent progress and a reality check. Aquat. Ecol. **32:** 261-279.

Sterner, R. W. and J. J. Elser. 2002. Ecological stoichiometry: the biology of elements from molecules to the biosphere, Princeton University Press.

Stevenson, R. J., M. L. Bothwell and R. L. Lowe. 1996. Algal ecology: freshwater benthic ecosystems, Academic Press.

Streit, B. 1981. Food searching and exploitation by a primary consumer (*Ancylus fluviatilis*) in a stochastic environment - nonrandom movement patterns. Rev. Suisse Zool. **88:** 887.

Swamikannu, X. and K. D. Hoagland. 1989. Effects of snail grazing on the diversity and structure of a periphyton community in a eutrophic pond. Can. J. Fish. Aquat. Sci. **46:** 1698-1704.

Tessier, A. J. and C. E. Goulden. 1982. Estimating food limitation in cladoceran populations. Limnol. Oceanogr. **27:** 707-717.

Thomas, J. D. 1986. The chemical ecology of *Biomphalaria glabrata* (Say): Sugars as attractants and arrestants. Comp. Biochem. Physiol. A **83:** 457-460.

Thomas, J. D., C. Cowley and J. Ofosu-Barko. 1980. Behavioural responses to amino acids and related compounds, including propionic acid, by adult *Biomphalaria glabrata* (Say), the snail host of *Schistosoma mansoni*. Comp. Biochem. Physiol. C **66:** 17-27.

Thomas, J. D., J. Ofosu-Barko and R. L. Patience. 1983. Behavioural responses to carboxylic and amino acids by *Biomphalaria glabrata* (Say), the snail host of *Schistosoma mansoni* (Sambon), and other freshwater molluscs. Comp. Biochem. Physiol. C **75:** 57-76.

Thomas, J. D., C. Kowalczyk and B. Somasundaram. 1989. The biochemical ecology of *Biomphalaria glabrata*, a snail host of *Schistosoma mansoni*: short chain carboxylic and amino acids as phagostimulants. Comp. Biochem. Physiol. A **93:** 899-911.

Thomas, J. D., P. R. Sterry, H. Jones, M. Gubala and B. M. Grealy. 1986. The chemical ecology of *Biomphalaria glabrata* (Say): Sugars as phagostimulants. Comp. Biochem. Physiol. A **83:** 461-475.

Tilman, D. 1982. Resource competition and community structure, Princeton University Press.

Turner, A. M., R. J. Bernot and C. M. Boes. 2000. Chemical cues modify species interactions: the ecological consequences of predator avoidance by freshwater snails. Oikos **88:** 148-158.

Urabe, J. and Y. Watanabe. 1992. Possibility of N or P limitation for planktonic cladocerans: An experimental test. Limnol. Oceanogr. **37:** 244-251.

Van Donk, E. 1989. The role of fungal parasites in phytoplankton succession. p. 171-194. *In* U. Sommer [eds.], Plankton Ecology. Springer.

Van Donk, E. and D. O. Hessen. 1993. Grazing resistance in nutrient-stressed phytoplankton. Oecologia **93:** 508-511.

Van Donk, E., M. Lürling, D. O. Hessen and G. M. Lokhorst. 1997. Altered cell wall morphology in nutrient-deficient phytoplankton and its impact on grazers. Limnol. Oceanogr. **42:** 357-364.

Van Gool, E. and J. Ringelberg. 1996. Daphnids respond to algal-associated odours. J. Plankton Res. **18:** 197-202.

Von Elert, E. 2002. Determination of limiting polyunsaturated fatty acids in *Daphnia galeata* using a new method to enrich food algae with single fatty acids. Limnol. Oceanogr. **47:** 1764-1773.

Von Elert, E. and C. J. Loose. 1996. Predator-induced diel vertical migration in *Daphnia*: enrichment and preliminary chemical characterization of a kairomone exuded by fish. J. Chem. Ecol. **22:** 885-895.

Von Elert, E. and P. Stampfl. 2000. Food quality for *Eudiaptomus gracilis*: the importance of particular highly unsaturated fatty acids. Freshw. Biol. **45:** 189-200.

Von Elert, E. and T. Wolffrom. 2001. Supplementation of cyanobacterial food with polyunsaturated fatty acids does not improve growth of *Daphnia*. Limnol. Oceanogr. **46:** 1552-1558.

Von Elert, E., D. Martin-Creuzburg and J. R. Le Coz. 2003. Absence of sterols constrains carbon transfer between cyanobacteria and a freshwater herbivore (*Daphnia galeata*). Proc. R. Soc. Lond. B. **270:** 1209-1214.

Wacker, A. 2002. Effects of biochemical food quality on the recruitment of *Dreissena polymorpha* in the littoral of Lake Constance: A field study and a laboratory approach. Ph.D. Thesis. University of Konstanz, Germany.

Wacker, A. and E. von Elert. 2001. Polyunsaturated fatty acids: evidence for non-substitutable biochemical resources in *Daphnia galeata*. Ecology **82:** 2507-2520.

Wacker, A. and E. von Elert. 2002. Strong influences of larval diet history on subsequent post-settlement growth in the freshwater mollusc *Dreissena polymorpha*. Proc. R. Soc. Lond. B. **269:** 2113-2119.

Wacker, A. and E. von Elert. 2003. Food quality controls reproduction of the zebra mussel (*Dreissena polymorpha*). Oecologia **135:** 332-338.

Wacker, A. and E. von Elert. 2004. Food quality controls egg quality of the zebra mussel *Dreissena polymorpha*: The role of fatty acids. Limnol. Oceanogr. **49:** 1794-1801.

Wacker, A., P. Becher and E. von Elert. 2002. Food quality effects of unsaturated fatty acids on larvae of the zebra mussel *Dreissena polymorpha*. Limnol. Oceanogr. **47:** 1242-1248.

Wakefield, R. L. and S. N. Murray. 1998. Factors influencing food choice by the seaweed-eating marine snail *Norrisia norrisi* (Trochidae). Mar. Biol. **130:** 631-642.

Watson, S. B. 2003. Cyanobacterial and eukaryotic algal odour compounds: signals or by-products? A review of their biological activity. Phycologia **42:** 332-350.

Watson, S. B. and J. Ridal. 2004. Periphyton: a primary source of widespread and severe taste and odour. Wat. Sci. Technol. **49:** 33-39.

Wendel, T. 1994. Lipoxygenase-katalysierte VOC-Bildung - Untersuchungen an Seewasser und Laborkulturen von Diatomeen. PhD Thesis. University of Tübingen, Germany.

Wendel, T. and F. Jüttner. 1996. Lipoxygenase-mediated formation of hydrocarbons and unsaturated aldehydes in freshwater diatoms. Phytochemistry **41:** 1445-1449.

Wessels, M. 1998. Geological history of the Lake Constance area. Arch. Hydrobiol. Spec. Issues Advanc. Limnol. **53:** 1-12.

Wetzel, R. G. 2001. Limnology, 3rd. Academic Press.

Wisenden, B. D. 2000. Olfactory asessment of predation risk in the aquatic environment. Phil. Trans. R. Soc. Lond. B **355:** 1205-1208.

Record of achievement / Abgrenzung der Eigenleistung

Chapter 2:	I contributed to design, field sampling and sample processing for the survey and exclusively analyzed the collected data.
Chapter 3, 4, 5 and 6:	Results described in these chapters were exclusively performed by myself or under my direct supervision.

Acknowledgements

Sincere thanks to everybody who supported me during the work on this thesis.

First of all, I want to thank my supervisor, PD Dr. Eric von Elert for proposing this topic and offering me a place in his working group, but especially for having an open door all the time for my questions. Thanks for the long and critical discussions that greatly improved my way of working in general and this thesis in particular. I also want to thank Prof. Dr. Karl-Otto Rothhaupt for the evaluation of the present study.

I am very grateful to Miquel Lürling who showed me how exciting research in aquatic ecology can be - without his contagious enthusiasm I would not be where I am today.

My fellow PhD students at the Limnological Institute made the last three years a good time. This was in particular, but not exclusively, due to skiing trips and numerous barbecues on the shores of Lake Constance. Thanks a lot! In particular, I want to thank my roommate Dominik Martin-Creuzburg for lots of time for discussions (scientific and many others), and the ‚people from next door', Astrid Löffler, Nicole Scheifhacken and Lars Peters for ☐D and much more.

The research divers group at the Limnological Institute is cordially thanked for several snail collection dives and for giving me the opportunity for a lot field work - it was fun most of the times!

Without my student assistants, this work would not have been possible. Sincere thanks to Timo Basen, Christoph Berron, Sven Boekhoff, Andreas Fibich, Benjamin Kumpfmüller, Sandra Leist and Oliver Miler.

Thanks to all the other people at the Limnological Institute. In particular to Petra Merkel for excellent support with many analyses of carbon, fatty acid and VOC samples. Not to forget Oliver Walenciak who did a great job in helping me while Petra was not at the Institute. I also want to express sincere thanks to Christine Gebauer for all the rapid and precise nutrient analyses and to Martin Wolf and Jürgen Gans-Thomsen for technical support and help with the building of experimental setups.

Thanks to the administration department of SFB 454, Mrs. S. Berger and especially Mrs. U. Tschakert, who relieved me of a lot of the bureaucracy.

Many thanks to Armin Glaschke for a last-minute check on the English of chapters 1 and 7.

Last but not least, I want to cordially thank my parents, who always supported me and Christa, who never lost the confidence that I would succeed and for everything else.

Research on this project was funded by the German Research Foundation (DFG) within the Collaborative Research Area SFB 454 - ‚Littoral of Lake Constance'.

Curriculum Vitae

Name	Patrick Fink
Day of birth	May 5th, 1976
Place of birth	Reutlingen
Nationality	German

Scientific career	since July, 2005 Post-doctoral research associate at the Max-Planck-Institute for Limnology, Department of Physiological Ecology, Plön, Germany.
	2001–2005 Doctoral thesis entitled 'Food quality and food choice in freshwater gastropods: Field and laboratory investigations on a key component of littoral food webs' at the Limnological Institute, University of Konstanz, supervisor PD Dr. Eric von Elert
	2001 Diploma thesis entitled 'Untersuchungen zur Coenobieninduktion in *Scenedesmus* durch *Daphnia*' at the Limnological Institute, University of Konstanz, supervision Prof. Dr. Karl-Otto Rothhaupt and PD Dr. Eric von Elert
University education	2001 Diploma-degree at the University of Konstanz, Germany
	1999–2000 Six months practical research experience at the Netherlands Institute for Ecology - Centre for Limnology at Nieuwersluis, The Netherlands in the workgroup Food Web Studies under supervision of Prof. Dr. Ellen van Donk and Dr. Miquel Lürling
	1998–2000 Studies of Limnology, Environmental Toxicology, Microbiology and Cell Biology at the Universitiy of Konstanz, Germany
	1998 Vordiplom degree in biology, University of Tübingen, Germany
	1996–1998 Basic studies of biology, University of Tübingen, Germany

List of Publications

PATRICK FINK and Eric Von Elert (*in prep.*) You are what you eat - Growth and fatty acid composition of the freshwater gastropod *Radix ovata* fed algae with natural and modified fatty acid patterns.

PATRICK FINK, Eric Von Elert, and Friedrich Jüttner (*submitted*) Aldehydes from freshwater diatoms act as attractants for a benthic herbivore. *Submitted to Arch. Hydrobiol.*

PATRICK FINK, Eric Von Elert, and Friedrich Jüttner (*submitted*) Volatile foraging kairomones in the littoral: attractance of algal odors for a herbivorous freshwater gastropod. *Submitted to J. Chem. Ecol.*

PATRICK FINK and Eric Von Elert (*submitted*) Stoichiometric constraints in benthic food webs: nutrient limitation and compensatory feeding in the freshwater snail *Radix ovata. Submitted to Oikos.*

PATRICK FINK, Lars Peters, and Eric Von Elert (*in press*) Stoichiometric mismatch between littoral invertebrates and their periphyton food. *Arch. Hydrobiol.*

PATRICK FINK, and Eric Von Elert (*in press*) Food quality of algae and cyanobacteria for the freshwater gastropod *Bithynia tentaculata*: the role of polyunsaturated fatty acids. *Verh. Int. Verein. Limnol. (SIL).* ***29**:*

PATRICK FINK und Eric Von Elert (*in press*) Das C:N:P-Verhältnis von Periphyton und seine ökologische Relevanz für herbivore Makroinvertebraten. *Tagungsbericht, Deutsche Gesellschaft für Limnologie (DGL) - Jahrestagung 2005 (Karlsruhe).*

PATRICK FINK und Eric Von Elert (2004) Freisetzung flüchtiger organischer Verbindungen aus benthischen Süßwasseralgen durch weidende Schnecken. *Tagungsbericht, Deutsche Gesellschaft für Limnologie (DGL) - Jahrestagung 2003 (Köln)* : 569-573.

PATRICK FINK und Eric Von Elert (2003) Chemosensorische Futterwahl limnischer Gastropoden. *Tagungsbericht, Deutsche Gesellschaft für Limnologie (DGL) - Jahrestagung 2002 (Braunschweig)*: 329-333.

PATRICK FINK und Eric Von Elert (2002) *Daphnia*-induzierte Coenobienbildung bei *Scenedesmus obliquus* - chemische Charakterisierung des Kairomons. *Tagungsbericht, Deutsche Gesellschaft für Limnologie (DGL) - Jahrestagung 2001 (Kiel): 579-584.*

Erik J. Van Hannen, PATRICK FINK, and Miquel Lürling (2002) A revised secondary structure model for the internal transcribed spacer 2 of the green algae *Scenedesmus* and *Desmodesmus* and its implications for the phylogeny of these algae. *European Journal of Phycology* ***37**: 203-208.*